Tucholsky Wagner Zola Scott Sydow Schlegel
Turgenev Wallace Fonatne Freud
Twain Walther von der Vogelweide Fouqué Friedrich II. von Preußen
Weber Freiligrath Frey
Fechner Kant Ernst
Fichte Weiße Rose von Fallersleben Richthofen Frommel
Fehrs Engels Fielding Hölderlin
Faber Flaubert Eichendorff Tacitus Dumas
Feuerbach Maximilian I. von Habsburg Fock Eliasberg Zweig Ebner Eschenbach
Ewald Eliot Vergil
Goethe Elisabeth von Österreich London
Mendelssohn Balzac Shakespeare Dostojewski Ganghofer
Trackl Lichtenberg Rathenau Doyle Gjellerup
Mommsen Stevenson Tolstoi Lenz Hambruch
Thoma von Arnim Hanrieder Droste-Hülshoff
Dach Verne Hägele Hauff Humboldt
Karrillon Reuter Rousseau Hagen Hauptmann Gautier
Garschin Defoe Hebbel Baudelaire
Damaschke Descartes
Wolfram von Eschenbach Hegel Kussmaul Herder
Bronner Darwin Dickens Schopenhauer Rilke George
Melville Grimm Jerome Bebel
Campe Horváth Aristoteles Voltaire Federer Proust
Bismarck Vigny Barlach Herodot
Gengenbach Heine
Storm Casanova Tersteegen Grillparzer Georgy
Chamberlain Lessing Langbein Gilm
Brentano Lafontaine Gryphius
Strachwitz Claudius Schiller Kralik Iffland Sokrates
Katharina II. von Rußland Bellamy Schilling
Gerstäcker Raabe Gibbon Tschechow
Löns Hesse Hoffmann Gogol Wilde Vulpius
Luther Heym Hofmannsthal Morgenstern Gleim
Roth Klee Hölty Goedicke
Heyse Klopstock Kleist
Luxemburg Puschkin Homer Mörike
Machiavelli La Roche Horaz Musil
Navarra Aurel Musset Kierkegaard Kraft Kraus
Nestroy Marie de France Lamprecht Kind Kirchhoff Hugo Moltke
Nietzsche Laotse Ipsen Liebknecht
Marx Nansen Ringelnatz
von Ossietzky Lassalle Gorki Klett Leibniz
May vom Stein Lawrence Irving
Petalozzi Knigge
Platon Pückler Michelangelo Kafka
Sachs Poe Kock
Liebermann Korolenko
de Sade Praetorius Mistral Zetkin

The publishing house tredition has created the series **TREDITION CLASSICS**. It contains classical literature works from over two thousand years. Most of these titles have been out of print and off the bookstore shelves for decades.

The book series is intended to preserve the cultural legacy and to promote the timeless works of classical literature. As a reader of a **TREDITION CLASSICS** book, the reader supports the mission to save many of the amazing works of world literature from oblivion.

The symbol of **TREDITION CLASSICS** is Johannes Gutenberg (1400 – 1468), the inventor of movable type printing.

With the series, tredition intends to make thousands of international literature classics available in printed format again – worldwide.

All books are available at book retailers worldwide in paperback and in hardcover. For more information please visit: www.tredition.com

tredition was established in 2006 by Sandra Latusseck and Soenke Schulz. Based in Hamburg, Germany, tredition offers publishing solutions to authors and publishing houses, combined with worldwide distribution of printed and digital book content. tredition is uniquely positioned to enable authors and publishing houses to create books on their own terms and without conventional manufacturing risks.

For more information please visit: www.tredition.com

Cactus Culture for Amateurs
Being Descriptions of the Various Cactuses Grown in This Country, With Full and Practical Instructions for Their Successful Cultivation

W. Watson

Imprint

This book is part of the TREDITION CLASSICS series.

Author: W. Watson
Cover design: toepferschumann, Berlin (Germany)

Publisher: tredition GmbH, Hamburg (Germany)
ISBN: 978-3-8491-7323-4

www.tredition.com
www.tredition.de

Copyright:
The content of this book is sourced from the public domain.

The intention of the TREDITION CLASSICS series is to make world literature in the public domain available in printed format. Literary enthusiasts and organizations worldwide have scanned and digitally edited the original texts. tredition has subsequently formatted and redesigned the content into a modern reading layout. Therefore, we cannot guarantee the exact reproduction of the original format of a particular historic edition. Please also note that no modifications have been made to the spelling, therefore it may differ from the orthography used today.

FIG. 1. – A **COLLECTION OF CACTUSES.** *Frontispiece*

PREFACE

HE idea that Cactuses were seldom seen in English gardens, because so little was known about their cultivation and management, suggested to the Publisher of this book that a series of chapters on the best kinds, and how to grow them successfully, would be useful. These chapters were written for and published in *The Bazaar,* in 1885 and following years. Some alterations and additions have been made, and the whole is now offered as a thoroughly practical and descriptive work on the subject.

The descriptions are as simple and complete as they could be made; the names here used are those adopted at Kew; and the cultural directions are as full and detailed as is necessary. No species or variety is omitted which is known to be in cultivation, or of sufficient interest to be introduced. The many excellent figures of Cactuses in the *Botanical Magazine* (Bot. Mag.) are referred to under each species described, except in those cases where a complete figure is given in this book. My claims to be heard as a teacher in this department are based on an experience of ten years in the care and cultivation of the large collection of Cactuses at Kew.

Whatever the shortcomings of my share of the work may be, I feel certain that the numerous and excellent illustrations which the Publisher has obtained for this book cannot fail to render it attractive, and, let us also hope, contribute something towards bringing Cactuses into favour with horticulturists, professional as well as amateur.

W. WATSON.

CONTENTS.

INTRODUCTION

BOTANICAL CHARACTERS

CULTIVATION

PROPAGATION

THE GENUS EPIPHYLLUM

THE GENUS PHYLLOCACTUS

THE GENUS CEREUS

THE GENUS ECHINOCACTUS

THE GENUS ECHINOPSIS

THE GENUS MELOCACTUS

THE GENUS PILOCEREUS

THE GENUS MAMILLARIA

THE GENUS LEUCHTENBERGIA

THE GENUS PELECYPHORA

THE GENUS OPUNTIA

THE GENUS PERESKIA

THE GENUS RHIPSALIS

TEMPERATURES

DEALERS IN CACTUSES

INDEX OF SPECIES

CACTUS CULTURE

FOR AMATEURS

CHAPTER I

INTRODUCTION.

THE Cactus family is not popular among English horticulturists in these days, scarcely half a dozen species out of about a thousand known being considered good enough to be included among favourite garden plants. Probably five hundred kinds have been, or are, in cultivation in the gardens of the few specialists who take an interest in Cactuses; but these are practically unknown in English horticulture. It is not, however, very many years ago that there was something like a Cactus mania, when rich amateurs vied with each other in procuring and growing large collections of the rarest and newest kinds.

"About the year 1830, Cacti began to be specially patronised by several rich plant amateurs, of whom may be mentioned the Duke of Bedford, who formed a fine collection at Woburn Abbey, the Duke of Devonshire, and Mr. Harris, of Kingsbury. Mr. Palmer, of Shakelwell, had become possessed of Mr. Haworth's collection, to which he greatly added by purchases; he, however, found his rival in the Rev. H. Williams, of Hendon, who formed a fine and select collection, and, on account of the eagerness of growers to obtain the new and rare plants, high prices were given for them, ten, twelve, and even twenty and thirty guineas often being given for single plants of the Echinocactus. Thus private collectors were induced to forward from their native countries—chiefly from Mexico and Chili—extensive collections of Cacti." (quoting J. Smith. A.L.S., ex-Curator of the Royal Gardens. Kew).

This reads like what might be written of the position held now in England by the Orchid family, and what has been written of Tulips and other plants whose popularity has been great at some time or other. Why have Cactuses gone out of favour? It is impossible to give any satisfactory answer to this question. No doubt they belong to that class of objects which is only popular whilst it pleases the eye or tickles the fancy; and the eye and the fancy having tired of it, look to something different.

The general belief with respect to Cactuses is that they are all wanting in beauty, that they are remarkable only in that they are exceedingly curious in form, and as a rule very ugly. It is true that none of them possess any claims to gracefulness of habit or elegance of foliage, such as are usual in popular plants, and, when not in flower, very few of the Cactuses would answer to our present ideas of beauty with respect to the plants we cultivate. Nevertheless, the stems of many of them (see Frontispiece, Fig. 1) are peculiarly attractive on account of their strange, even fantastic, forms, their spiny clothing, the absence of leaves, except in very few cases, and their singular manner of growth. To the few who care for Cactuses there is a great deal of beauty, even in these characters, although perhaps the eye has to be educated up to it.

If the stems are more curious than beautiful, the flowers of the majority of the species of Cactuses are unsurpassed, as regards size and form, and brilliancy and variety in colour, by any other family of plants, not even excluding Orchids. In size some of the flowers equal those of the Queen of Water Lilies *(Victoria regia),* whilst the colours vary from the purest white to brilliant crimson and deep yellow. Some of them are also deliciously fragrant. Those kinds which expand their huge blossoms only at night are particularly interesting; and in the early days of Cactus culture the flowering of one of these was a great event in English gardens.

Of the many collections of Cactuses formed many years ago in England, that at Kew is the only one that still exists. This collection has always been rich in the number of species it contained; at the present time the number of kinds cultivated there is about 500. Mr. Peacock, of Hammersmith, also has a large collection of Cactuses, many of which he has at various times exhibited in public places, such as the Crystal Palace, and the large conservatory attached to the Royal Horticultural Society's Gardens at South Kensington. Other smaller collections are cultivated in the Botanic Gardens at Oxford, Cambridge, Glasnevin, and Edinburgh.

A great point in favour of the plants of the Cactus family for gardens of small size, and even for window gardening—a modest phase of plant culture which has made much progress in recent years—is the simpleness of their requirements under cultivation. No plants give so much pleasure in return for so small an amount of attention as do these. Their peculiarly tough-skinned succulent stems enable them to go for an extraordinary length of time without water; indeed, it may be said that the treatment most suitable for many of them during the greater portion of the year is such as would be fatal to most other plants. Cactuses are children of the dry barren plains and mountain sides, living where scarcely any other form of vegetation could find nourishment, and thriving with the scorching heat of the sun over their heads, and their roots buried in the dry, hungry soil, or rocks which afford them anchorage and food.

In beauty and variety of flowers, in the remarkable forms of their stems, in the simple nature of their requirements, and in the other points of special interest which characterise this family, and which supply the cultivator and student with an unfailing source of pleasure and instruction, the Cactus family is peculiarly rich.

CHAPTER II.

BOTANICAL CHARACTERS.

LTHOUGH strictly botanical information may be considered as falling outside the limits of a treatise intended only for the cultivator, yet a short account of the principal characters by which Cactuses are grouped and classified may not be without interest.

From the singular form and succulent nature of the whole of the Cactus family, it might be inferred that, in these characters alone, we have reliable marks of relationship, and that it would be safe to call all those plants Cactuses in which such characters are manifest. A glance at some members of other families will, however, soon show how easily one might thus be mistaken. In the Euphorbias we find a number of kinds, especially amongst those which inhabit the dry, sandy plains of South Africa, which bear a striking resemblance to many of the Cactuses, particularly the columnar ones and the Rhipsalis. (The Euphorbias all have milk-like sap, which, on pricking their stems or leaves, at once exudes and thus reveals their true character. The sap of the Cactuses is watery). Amongst Stapelias, too, we meet with plants which mimic the stem characters of some of the smaller kinds of Cactus. Again, in the Cactuses themselves we have curious cases of plant mimicry; as, for instance, the Rhipsalis, which looks like a bunch of Mistletoe, and the Pereskia, the leaves and habit of which are more like what belong to, say, the Gooseberry family than to a form of Cactus. From this it will be seen that although these plants are almost all succulent, and curiously formed, they are by no means singular in this respect.

The characters of the order are thus defined by botanists: Cactuses are either herbs, shrubs, or trees, with soft flesh and copious watery juice. Root woody, branching, with soft bark. Stem branching or simple, round, angular, channelled, winged, flattened, or cylindrical; sometimes clothed with numerous tufts of spines which vary in texture, size, and form very considerably; or, when spineless, the stems bear numerous dot-like scars, termed areoles. Leaves very minute, or entirely absent, falling off very early, except in the Pereskia and several of the Opuntias, in which they are large, fleshy, and persistent. Flowers solitary, except in the Pereskia, and borne on the top or side of the stem; they are composed of numerous parts or segments; the sepals and petals are not easily distinguished from each other; the calyx tube is joined to, or combined, with the ovary, and is often covered with scale-like sepals and hairs or spines; the calyx is sometimes partly united so as to form a tube, and the petals are spread in regular whorls, except in the Epiphyllum. Stamens many, springing from the side of the tube or throat of the calyx, sometimes joined to the petals, generally equal in length; anthers small and oblong. Ovary smooth, or covered with scales and spines, or woolly, one-celled; style simple, filiform or cylindrical, with a stigma of two or more spreading rays, upon which are small papillae. Fruit pulpy, smooth, scaly, or spiny, the pulp soft and juicy, sweet or acid, and full of numerous small, usually black, seeds.

Tribe I.—*Calyx tube produced beyond the Ovary. Stem covered with Tubercles, or Ribs, bearing Spines.*

1. MELOCACTUS. Stem globose; flowers in a dense cap-like head, composed of layers of bristly wool and slender spines, amongst which the small flowers are developed. The cap is persistent, and increases annually with the stem.

2. MAMILLARIA. Stems short, usually globose, and covered with tubercles or mammae, rarely ridged, the apex bearing spiny cushions; flowers mostly in rings round the stem.

3. PELECYPHORA. Stem small, club-shaped; tubercles in spiral rows, and flattened on the top, where are two rows of short scale-like spines.

4. LEUCHTENBERGIA. Stem naked at the base; tubercles on the upper part large, fleshy, elongated, three-angled, bearing at the apex a tuft of long, thin, gristle-like spines.

5. ECHINOCACTUS. Stem short, ridged, spiny; calyx tube of the flower large, bell-shaped; ovary and fruit scaly.

6. DISCOCACTUS. Stem short; calyx tube thin, the throat filled by the stamens; ovary and fruit smooth.

7. CEREUS. Stem often long and erect, sometimes scandent, branching, ridged or angular; flowers from the sides of the stem; calyx tube elongated and regular; stamens free.

8. PHYLLOCACTUS. Stem flattened, jointed, and notched; flowers from the sides, large, having long, thin tubes and a regular arrangement of the petals.

9. EPIPHYLLUM. Stem flattened, jointed; joints short; flowers from the apices of the joints; calyx tube short; petals irregular, almost bilabiate.

Tribe II.—*Calyx-tube not produced beyond the Ovary. Stem branching, jointed.*

10. RHIPSALIS. Stem thin and rounded, angular, or flattened, bearing tufts of hair when young; flowers small; petals spreading; ovary smooth; fruit a small pea-like berry.

11. OPUNTIA. Stem jointed, joints broad and fleshy, or rounded; spines barbed; flowers large; fruit spinous, large, pear-like.

12. PERESKIA. Stem woody, spiny, branching freely; leaves fleshy, large, persistent; flowers medium in size, in panicles on the ends of the branches.

The above is a key to the genera on the plan of the most recent botanical arrangement, but for horticultural purposes it is necessary that the two genera Echinopsis and Pilocereus should be kept up. They come next to Cereus, and are distinguished as follows:

ECHINOPSIS. Stem as in Echinocactus, but the flowers are produced low down from the side of the stem, and the flower tube is long and curved.

PILOCEREUS. Stem tall, columnar, bearing long silky hairs as well as spines; flowers in a head on the top of the stem, rarely produced.

With the aid of this key anyone ought to be able to make out to what genus a particular Cactus belongs, and by referring to the descriptions of the species, he may succeed in making out what the plant is.

For the classification of Cactuses, botanists rely mainly on their floral organs and fruit. We may, therefore, take a plant of Phyllocactus, with which most of us are familiar, and, by observing the structure of its flowers, obtain some idea of the botanical characters of the whole order.

Phyllocactus has thin woody stems and branches composed of numerous long leaf-like joints, growing out of one another, and resembling thick leaves joined by their ends. Along the sides of these joints there are numerous notches, springing from which are the large handsome flowers. On looking carefully, we perceive that the long stalk-like expansion is not a stalk, because it is above the seed vessel, which is, of course, a portion of the flower itself. It is a hollow tube, and contains the long style or connection between the seed vessel and the stigma, a (Fig. 2). This tube, then, must be the calyx, and the small scattered scale-like bodies, b (Fig. 2), which clothe the outside, are really calyx lobes.

FIG. 2.—FLOWER OF PHYLLOCACTUS, CUT LENGTHWISE

a, Calyx Tube. *b,* Calyx Lobes. *c,* Ditto, assuming the form of Petals. *d,* Stamens. *e,* Style. *f,* Ovary or Seed Vessel.

Nearer the top of the flower, these calyx lobes are better developed, until, surrounding the corolla, we find them assuming the form and appearance of petals, *c* (Fig. 2). The corolla is composed of a large number of long strap-shaped pointed petals, very thin and delicate, often beautifully coloured, and generally spreading outwards. Springing from the bases of these petals, we find the stamens, *d* (Fig. 2), a great number of them, forming a bunch of threads unequal in length, and bearing on their tips the hay-seed-like anthers, which are attached to the threads by one of their points. The style is a long cylindrical body, *e* (Fig. 2), which stretches from the ovary to the top of the flower, where it splits into a head of spreading linear rays, = in. length. When the flower withers, the seed vessel, *f* (Fig. 2), remains on the plant and expands into a large succulent fruit, inside which is a mass of pulpy matter, inclosing the numerous, small, black, bony seeds.

It must not be supposed that all the genera into which Cactuses are divided are characterised by large flowers such as would render their study as easy as the genus taken as an illustration. In some, such for instance as the Rhipsalis, the flowers are small, and therefore less easy to dissect than those of Phyllocactus.

The stems of Cactuses show a very wide range of variation in size, in form, and in structure. In size, we have the colossal *Cereus giganteus,* whose straight stems when old are as firm as iron, and rise with many ascending arms or rear their tall leafless trunks like ships' masts to a height of 60 ft. or 70 ft. From this we descend through a multitude of various shapes and sizes to the tiny tufted Mamillarias, no larger than a lady's thimble, or the creeping Rhipsalis, which lies along the hard ground on which it grows, and looks like hairy caterpillars. In form, the variety is very remarkable. We have the Mistletoe Cactus, with the appearance of a bunch of Mistletoe, berries and all; the Thimble Cactus; the Dumpling Cactus; the Melon Cactus; the Turk's cap Cactus; the Rat's-tail Cactus; the Hedgehog Cactus; all having a resemblance to the things whose names they bear. Then there is the Indian Fig, with branches like battledores, joined by their ends; the Epiphyllum and Phyllocactus, with flattened leaf-like stems; the columnar spiny Cereus, with deeply channelled stems and the appearance of immense candelabra. Totally devoid of leaves, and often skeleton-like in appearance, these plants have a strange look about them, which is suggestive of some fossilised forms of vegetation belonging to the past ages of the mastodon, the elk, and the dodo, rather than to the living things of to-day.

By far the greater part of the species of Cactuses belong to the group with tall or elongated stems. "It is worthy of remark that as the stems advance in age the angles fill up, or the articulations disappear, in consequence of the slow growth of the woody axis and the gradual development of the cellular substance; so that, at the end of a number of years, all the branches of Cactuses, however angular or compressed they originally may have been, become trunks that are either perfectly cylindrical, or which have scarcely any visible angles."

A second large group is that of which the Melon and Hedgehog Cactuses are good representatives, which have sphere-shaped stems, covered with stout spines. We have hitherto spoken of the Cactuses as being without leaves, but this is only true of them when in an old or fully-developed state. On many of the stems we find upon their surface, or angles, small tubercles, which, when young, bear tiny scale-like leaves. These, however, soon wither and fall off, so that, to all appearance, leaves are never present on these plants. There is one exception, however, in the Barbadoes Gooseberry (Pereskia), which bears true and persistent leaves; but these may be considered anomalous in the order.

The term "succulent" is applied to Cactuses because of the large proportion of cellular tissue, *i.e.*, flesh, of their stems, as compared with the woody portion. In some of them, when young, the woody system appears to be altogether absent, and they have the appearance of a mass of fleshy matter, like a vegetable marrow. This succulent mass is protected by a tough skin, often of leather-like firmness, and almost without the little perforations called breathing and evaporating pores, which in other plants are very numerous. This enables the Cactuses to sustain without suffering the full ardour of the burning sun and parched-up nature of the soil peculiar to the countries where they are native. Nature has endowed Cactuses with a skin similar to what she clothes many succulent fruits with, such as the Apple, Plum, Peach, &c., to which the sun's powerful rays are necessary for their growth and ripening.

The spiny coat of the majority of Cactuses is no doubt intended to serve as a protection from the wild animals inhabiting with them the sterile plains of America, and to whom the cool watery flesh of the Cactus would otherwise fall a prey. Indeed, these spines are not sufficient to prevent some animals from obtaining the watery insides of these plants, for we read that mules and wild horses kick them open and greedily devour their succulent flesh. It has also been suggested that the spines are intended to serve the plants as a sort of shade from the powerful sunshine, as they often spread over and interlace about the stems.

CHAPTER III.

CULTIVATION.

Y noting the conditions in which plants are found growing in a natural state, we obtain some clue to their successful management, when placed under conditions more or less artificial; and, in the case of Cactuses, knowledge of this kind is of more than ordinary importance. In the knowledge that, with only one or two exceptions, they will not exist in any but sunny lands, where, during the greater part of the year, dry weather prevails, we perceive what conditions are likely to suit them when under cultivation in our plant-houses.

Cactuses are all American (using this term for the whole of the New World) with only one or two exceptions (several species of Rhipsalis have been found wild in Africa, Madagascar, and Ceylon), and, broadly speaking, they are mostly tropical plants, notwithstanding the fact of their extending to the snow-line on some of the Andean Mountains of Chili, where several species of the Hedgehog Cactus were found by Humboldt on the summit of rocks whose bases were planted in snow. In California, in Mexico and Texas, in the provinces of Central and South America, as far south as Chili, and in many of the islands contiguous to the mainland, the Cactus family has become established wherever warmth and drought, such as its members delight in, allowed them to get established. In many of the coast lands, they occur in very large numbers, forming forests of strange aspect, and giving to the landscape a weird, picturesque appearance. Humboldt, in his "Views of Nature," says: "There is hardly any physiognomical character of exotic vegetation that produces a more singular and ineffaceable impression on the mind of the traveller than an arid plain, densely covered with columnar or candelabra-like stems of Cactuses, similar to those near Cumana, New Barcelona, Cora. and in the province of Jaen de Bracamoros." This applies also to some of the small islands of the West Indies, the hills or mountains of which are crowned with these curious-looking plants, whose singular shapes are alone sufficient to remind the traveller that he has reached an American coast; for these Cactuses are as peculiar a feature of the New World as the Heaths are in the Old, or as Eucalypti are in Australia.

Although the Cactus order is, in its distribution by Nature, limited to the regions of America, yet it is now represented in various parts of the Old World by plants which are apparently as wild and as much at home as when in their native countries.

The Indian Figs are, perhaps, the most widely distributed of Cactuses in the Old World—a circumstance due to their having been introduced for the sake of their edible fruits, and more especially for the cultivation of the cochineal insect. In various places along the shores of the Mediterranean, and in South Africa, and even in Australia, the Opuntias have become naturalised, and appear like aboriginal inhabitants. It is, however, only in warm sunny regions that the naturalisation of these plants is possible.

From these facts, we are able to form some general idea of the conditions suitable for Cactuses when cultivated in our greenhouses; for, although we seldom have, or care to have, any but diminutive specimens of many of these plants as compared with their appearance when wild, yet we know that the same conditions as regards heat, light, and moisture are necessary for small Cactuses as for full-grown ones.

Although the places in which Cactuses naturally abound are, for the greater portion of the year, very dry and warm, heavy rains are more or less frequent during certain periods, and these, often accompanied by extreme warmth and bright sunshine, have an invigorating and almost forcing effect on the growth of Cactuses. It is during this rainy period that the whole of the growth is made, and new life is, as it were, given to the plant, its reservoir-like structure enabling it to store up a large amount of food and moisture, so that on the return of dry weather the safety of the plant is insured.

It is to the management of Cactuses in a small state, such as is most convenient for our plant-houses, and not to the cultivation of those colossal species referred to above, that the instructions given here will be for the most part devoted; but, as in the case of almost every one of our cultivated plants, it is important to the cultivator to know something of the conditions which Nature has provided for Cactuses in those lands where they are native.

There is nothing in the nature or the requirements of Cactuses that should render their successful management beyond the means of anyone who possesses a small, heated greenhouse, or even a window recess to which sunlight can be admitted during some portion of the day. In large establishments, such as Kew, it is possible to provide a spacious house specially for the cultivation of an extensive collection, where many of them may attain a good size before becoming too big. And it will be evident that where a house such as that at Kew can be afforded, much more satisfactory results may generally be obtained, than if plants have to be provided for in a house containing various other plants, or in the window of a dwelling-room. Apart altogether from size, it is, however, possible to grow a collection of Cactuses, and to grow them well, in a house of small dimensions—given the amount of sunlight and heat which are required by these plants. We sometimes see Cactuses—specimens, too, of choice and rare kinds—which have been reared in a cottager's window or in a small greenhouse, and which in health and beauty have at least equalled what has been accomplished in the most elaborately prepared houses. It may be said that these successes, under conditions of the most limited kind, are accidental rather than the result of properly understood treatment; but however they have been brought about, these instances of good cultivation are sufficient to show that success is possible, even where the means are of the simplest or most restricted kind. Whether it be in a large house, fitted with the best arrangements, or in the window of the cottager, the conditions essential to the successful cultivation of Cactuses are practically the same.

In Wardian Cases.—Many of our readers will be acquainted with the neat little glass cases, like greenhouses in shape, and fitted up in much the same way, which are sometimes to be seen in our markets, filled with a collection of miniature Cactuses. To the professional gardener, these cases are playthings, and are looked upon by him as bearing about the same relation to gardening as a child's doll's house does to housekeeping. Not-withstanding this, they are the source of much interest, and even of instruction, to many of the millions to whom a greenhouse or serious gardening is an impossibility. In these little cases—for which we are indebted to Mr. Boller, a dealer in Cactaceous plants—it is possible to grow a collection of tiny Cactuses for years, if only the operations of watering, potting, ventilating, and other matters connected with ordinary plant growing, are properly attended to.

In Window Recesses.—In the window recess larger specimens may be grown, and here it is possible to grow and flower successfully many of the plants of the Cactus family. In a window with a south aspect, and which lights a room where fires are kept, at least during cold weather, specimens of Phyllocactus, *Cereus flagelliformis,* Epiphyllum, and, in fact, of almost every kind of Cactus, are sometimes to be met with even in England; whilst in Germany they are as popular among the poorer classes as the Fuchsia, the Pelargonium, and the Musk are with us. One of the commonest of Cactuses in the latter country is the Rat's-tail Cactus *(Cereus flagelliformis),* and it is no unusual thing to see a large window of a cottager's dwelling thickly draped on the inside with the long, tail-like growths and handsome rose-coloured flowers of this plant. This is only one among dozens of species, all equally useful for window gardening, and all as interesting and beautiful as those above described.

In Greenhouses.—For the greenhouse proper, Cactuses are well adapted, either as the sole occupants or as suitable for such positions as are afforded by shelves or baskets placed near the roof glass. If the greenhouse is not fitted with heating arrangements, then, by selecting only those species of Cactus that are known to thrive in a position where, during winter, they are kept safe out of the reach of frost (of which a large number are known) a good collection of these plants may be grown. In heated structures the selection of kinds may be made according to the space available, and to the conditions under which they will be expected to grow. Fig. 3 represents a section of a house for Cactuses, which will afford a good idea of the kind of structure best suited for them. The aspect is due south.

FIG. 3.—SECTION OF HOUSE FOR CACTUSES—A,A, Hot-water Pipes; B,B, Ventilators

When grown on their own roots, the Epiphyllums, as well as the pendent-growing kinds of Rhipsalis, and several species of Cereus, may be placed in baskets and suspended from the roof. The baskets should be lined with thin slices of fibrous peat, and the whole of the middle filled with the compost recommended for these plants under "Soil". When well managed, some very pretty objects are formed by the Epiphyllums grown as basket plants. The climbing Cactuses are usually planted in a little mound composed of loam and brick rubble, and their stems either trained along rafters or allowed to run up the back wall of a greenhouse, against which they root freely, and are generally capable of taking care of themselves with very little attention from the gardener.

In Frames.—For cultivation in frames, the conditions are the same as for greenhouses. Even when grown in the latter, it will be found conducive to the health and flowering of the plants if, during the summer months, they can be placed in a frame with a south aspect, removing them back to the house again on the decline of summer weather. Wherever the place selected for Cactuses may be, whether in a large plant-house, or a frame, or a window, it is of vital importance to the plants that the position should be exposed to bright sunshine during most of the day. Without sunlight, they can no more thrive than a Pelargonium could without water. In Germany, many growers of almost all the kinds of Cactuses place their young plants in frames, which are prepared as follows: In April or May a hot-bed of manure and leaves is prepared, and a frame placed upon it, looking south. Six inches of soil is put on the top of the bed, and in this, as soon as the temperature of the bed has fallen to about 70 deg., the young plants are placed in rows. The frames are kept close even in bright weather, except when there is too much moisture inside, and the plants are syringed twice daily in dry, hot weather. The growth they make under this treatment is astonishing. By the autumn the plants are ready to be ripened by exposure to sun and air, and in September they are lifted, planted in pots, and sent to market for sale. This method may be adopted in England, and if carefully managed, the growth the plants would make would far exceed anything ever accomplished when they are kept permanently in pots.

Out-of-doors.—There are some kinds which may be grown out of doors altogether, if planted on a sunny, sheltered position, on a rockery. The most successful plan is that followed at Kew, where a collection of the hardier species is planted in a rockery composed of brick rubble and stones. During summer the plants are exposed; but when cold weather and rains come, lights are placed permanently over the rockery, and in this way it is kept comparatively dry. No fire-heat or protection of any other kind is used, and the vigorous growth, robust health, and floriferousness of the several species are proofs of the fitness of the treatment for this class of plants.

In any garden where a few square yards in a sunny, well-drained position can be afforded for a raised rockery, the hardy Cactuses may be easily managed. To make a suitable rockery, proceed as follows: Find a position against the south wall of a house, greenhouse, or shed, and against this wall construct a raised rockery of brick rubble, lime rubbish, stones (soft sandstone, if possible), and fibrous loam. The rockery when finished should be, say, 4 ft. wide, and reach along the wall as far as required; the back of the rockery would extend about 2 ft. above the ground level, and fall towards the front. Fix in the wall, 1 ft. or so above the rockery, a number of hooks at intervals all along, to hold in position lights sufficiently long to cover the rockery from the wall to the front, where they could be supported by short posts driven in the ground. The lights should be removed during summer to some shed, and brought out for use on the approach of winter. Treated in this manner, the following hardy species could not fail to be a success:

Opuntia Rafinesquii and var. *arkansana, O. vulgaris, O. brachyarthra, O. Picolominiana, O. missouriensis, O. humilis, Cereus Fendleri, C. Engelmanni, C. gonacanthus, C. phoeniceus, Echinocactus Simpsoni, E. Pentlandii, Mamillaria vivipara.*

Having briefly pointed out the various positions in which Cactuses may be cultivated successfully, we will now proceed to treat in detail the various operations which are considered as being of more or less importance in their management. These are potting, watering, and temperatures, after which propagation by means of seeds, cuttings, and grafting, hybridisation, seed saving, &c., and diseases and noxious insects will be treated upon.

Soil.—The conditions in which plants grow naturally, are what we usually try to imitate for their cultivation artificially. At all events, such is supposed to be theoretically right, however difficult we may often find it to be in practice. Soil in some form or other is necessary to the healthy existence of all plants; and we know that the nature of the soil varies with that of the plants growing in it, or, in other words, certain soils are necessary to certain plants, whether in a state of nature or cultivated in gardens. But, whilst admitting that Nature, when intelligently followed, would not lead us far astray, we must be careful not to follow her too strictly when dealing with the management of plants in gardens. There are other circumstances besides the nature of the soil by which plants are influenced. Soil is only one of the conditions on which plants depend, and where the other conditions are not exactly the same in our gardens as in nature, it is often found necessary to employ a different soil from that in which the plants grow when wild.

It has been stated that plants do not grow naturally in the soil best suited for them, and that the reason why many plants are found in peculiar places is not at all because they prefer them, but because they alone are capable of existing there, or because they take refuge there from the inroads of stouter neighbours who would destroy them or crowd them out. There are, as every gardener knows, numerous plants that succeed equally well in widely different soils, and a soil which may be suitable for a plant in one place, may prove totally unsuited in another. Hence it is why we find one gardener recommending one kind of soil, and another a different one, for the same plant, both answering equally well because of other conditions fitting better with each soil. This helps us to understand how it is that many garden subjects grow much better when planted in composts often quite different from those the plants are found in when wild. Few plants have a particular predilection for soil, and some have what we may call the power to adapt themselves to conditions often widely different.

In Cactuses we have a family of plants for which special conditions are necessary; and, as regards soil, whether we are guided by nature or by gardening experience, we are led to conclude that almost all of them thrive only when planted in one kind, that soil being principally loam. Plants which are limited in nature to sandy, sun-scorched plains or the glaring sides of rocky hills and mountains, where scarcely any other form of vegetation can exist, are not likely to require much decayed vegetable humus, but must obtain their food from inorganic substances, such as loam, sand, or lime. So it is with them when grown in our houses. They are healthiest and longest-lived when planted in a loamy soil; and although they may be grown fairly well for a time when placed in a compost of loam and leaf mould, or loam and peat, yet the growth they make is generally too sappy and weak; it is simply fat without bone, which, when the necessary resting period comes round, either rots or gradually dries up. In preparing soil, therefore, for all Cactuses (except Epiphyllum and Rhipsalis, which will be treated separately) a good, rather stiff loam, with plenty of grass fibre in it, should form the principal ingredient, sand and, if obtainable, small brick rubble being added—one part of each of the latter to six parts of the former. The brick rubble should be pounded up so that the largest pieces are about the size of hazel nuts. Lime rubbish, *i.e.,* old plaster from buildings, &c., is sometimes recommended for Cactuses, but it does not appear to be of any use except as drainage. At Kew its use has been discontinued, and it is now generally condemned by all good cultivators. Of course, the idea that lime was beneficial to Cactuses sprang from the knowledge that it existed in large quantities in the soil in which the plants grew naturally, and it is often found in abundance, in the form of oxalate of lime, in the old stems of the plants. But in good loam, lime, in the state of chalk, is always present, and this, together with the lime contained in the brick rubble, is sufficient to supply the plants with as much as they require.

For Epiphyllums and Rhipsalis, both of which are epiphytal naturally, but which are found to thrive best in pots in our houses, a mixture of equal parts of peat and loam with sand and brick rubble in the same proportion as before recommended, will be found most suitable. Leaf mould is sometimes used for these plants; but unless really good it is best left out of the soil. The finest Epiphyllums have been grown in a soil which consists almost wholly of a light fibry loam, with the addition of a little crushed bones.

Potting.—Cactuses, when healthy, are injuriously affected by frequent disturbance at the roots. On the arrival of the potting season, which for these plants is in April and May, established plants should be examined at the root, and if the roots are found to be in a healthy condition, and the soil sweet, they should be replaced in the same pots to continue in them another year. If the roots are decayed, or the soil has become sour, it should be shaken away from the roots, which must be examined, cutting away all decayed portions, and shortening the longest roots to within a few inches of the base of the plant. Cactuses are so tenacious of life, and appear to rely so little on their roots, that it will be found the wisest plan, when re-potting them, to cut the roots thoroughly.

The size of pots most suitable is what would be considered small in comparison with other plants, Cactuses preferring to be somewhat cramped in this respect. This, indeed, is how they are found when wild, the roots generally fixing themselves in the crevices of the rocks or stones about which the plants grow, so that a large specimen is often found to have only a few inches of space in the cleft of a rock for the whole of its roots. When thus limited, growth is firmer and the flowers are produced in much greater profusion than when a liberal amount of root space is afforded. The pots should be well drained-about one-fifth of their depth filled with drainage when intended for large, strong-growing kinds, and one-third for the smaller ones, such as Mamillarias. A layer of rough fibry material should be placed over the crocks to prevent the finer soil from stopping the drainage. When filling in the soil, press it down firmly, spreading the roots well amongst it, and keeping the base of the plant only an inch or so below the surface.

For plants with weak stems, stakes will be necessary, and even stout-stemmed kinds, when their roots are not sufficient to hold them firmly, will do best if fastened to one or two strong stakes till they have made new roots and got firm hold of the soil. Epiphyllums, when grown as standards, should be tied to strong wire supports, those with three short, prong-like legs being most desirable, as, owing to the weight of the head of the plant, a single stake is not sufficient to hold the whole firmly. After potting, no water should be given for a few weeks. In fact, if the atmosphere in which the plants are placed be kept a little moist, it will not be necessary to water them till signs of fresh growth are perceived. For Epiphyllums and Rhipsalis, water will be required earlier than this; but even they are best left for a few days without water, after they have been repotted. As soon as fresh growth is perceived, the plants may be well watered, and from this time water may be supplied as often as the soil approaches dryness. Newly-imported plants, which on arrival are usually much shrivelled and rootless, should be potted in rather dry soil and small pots, and treated as recommended above. Cactuses, we must remember, contain an abundance of nourishment stored up in their stems, and upon this they will continue to exist for a considerable time without suffering; and, when their growing season comes round, root action commences whether the soil is wet or dry, the latter being the most favourable.

Plants altogether exposed to the air will push roots in due time. A remarkable instance of this has been recorded by Mr. J. R. Jackson, curator of the museums at Kew. A plant of *Pilocereus senilis,* which had grown too tall for the house, was cut off at the base, and placed in the museum as a specimen. Here it gradually dried up to within 2 ft. of the top, where a fracture across the stem had been made. Above this the stem remained fresh and healthy, and, on examining it some months afterwards, it was found that not only had the top of the stem remained green, but it had formed roots of its own, which had grown down the dead lower portion of the stem, and were in a perfectly healthy state. When it is remembered that all this happened in the dry atmosphere of a museum, it will be apparent how exceptional Cactuses are in their manner of growth, and in the wonderful tenacity of life they exhibit under conditions which would destroy the majority of plants in a very short time. We sometimes find, when examining the bases of Cactus stems, that decay has commenced; this is carefully cut out with a sharp knife, and the wound exposed to the action of the air till it is perfectly dry, or, as we term it, "callused."

Watering.—It will have peen gathered from what has been previously said in relation to the conditions under which the majority of the plants of the Cactus family grow when wild, that during their season of growth they require a good supply of moisture, both at the root and overhead; and afterwards a somewhat lengthened period of rest, that is, almost total dryness, accompanied by all the sunlight possible, and generally a somewhat high temperature. The growing season for all those kinds which require to be kept dry when at rest is from the end of April to the middle of August, and during this time they should be kept moderately moist, but not constantly saturated, which, however, is not likely to occur if the water is not carelessly supplied, and the drainage and soil are perfect. This treatment corresponds with what happens to Cactuses in a wild state, the frequent and heavy rains which occur in the earlier part of the summer in the American plains supplying the amount of moisture necessary to enable these plants to make fresh growth, and produce their beautiful flowers and spine-clothed fruits. After August, little or no rain falls, and the Cactuses assume a rather shrivelled appearance, which gives them an unhealthy look, but which is really a sign of ripeness, promising a plentiful crop of flowers when the rainy season again returns.

As the sun in England is not nearly so powerful as in the hot plains of Central America and the Southern States of North America, where Cactuses are found in greatest abundance, it will be evident that, if flowers are to be produced, we must see that our plants have a sufficiency of water in early summer, and little or none during the autumn and winter, whilst the whole year round they should be exposed to all the sunlight possible, the temperature, of course, varying with the requirements of the species, whether it is a native of tropical or of temperate regions. It is important that the cultivator should understand that if water is liberally supplied all through the summer, the plants cannot obtain the rest which is necessary to their ripening and producing flowers, as dryness at the root alone is not sufficient to provide this, but must be accompanied by exposure to bright sunlight, which is not possible in England during winter, so that the ripening process must begin before the summer is over.

It is possible to preserve most Cactuses alive by keeping them constantly growing; but, with very few exceptions, such treatment prevents the plants from flowering. The following is what is practised in the gardens where Cactuses are successfully cultivated. For the genera Cereus, Echinopsis, Echinocactus, Mamillaria, Opuntia, and Melocactus, a moist tropical house is provided, and in April the plants are freely watered at the root, and syringed overhead both morning and afternoon on all bright days. This treatment is continued till the end of July, when syringing is suspended, and the water supplied to the roots gradually reduced. By the end of August, the plants are placed in a large light frame with a south aspect, except the tall-growing kinds, which are too bulky to remove. In this frame the plants are kept till the summer is over, and are watered only about once a week should the sun be very powerful. The lights are removed on all bright sunny days, but are kept on during wet or dull weather, and at night. Under this treatment, many of the species assume a reddish appearance, and the thick fleshy-stemmed kinds generally shrivel somewhat. There is no occasion for alarm in the coloured and shrivelled appearance of the plants: on the contrary, it may be hailed as a good sign for flowers.

A common complaint in relation to Cacti as flowering plants is that they grow all right but rarely or never flower. The explanation of this is shown by the fact that the plants must be properly ripened and rested before they can produce flowers. On the approach of cold weather the plants which were removed to a frame to be ripened should be brought back into the house for the winter, and kept quite dry at the roots till the return of spring, when their flowers will be developed either before or soon after the watering season again commences.

Hitherto we have been dealing with those genera which have thick fleshy stems; but there still remain the genera Rhipsalis, Epiphyllum, and Phyllocactus, which are not capable of bearing the long period of drought advised for the former. The last-mentioned genus should, however, be kept almost dry at the root during winter, and, if placed in a light, airy house till the turn of the year, the branches will ripen, and set their flower buds much more readily than when they are wintered in a moist, partially-shaded house. During summer all the Phyllocactuses delight in plenty of water, and, when growing freely, a weak solution of manure affords them good food. Epiphyllums must be kept always more or less moist at the root, though, of course, when growing freely, they require more water than when growth has ceased for the year, which happens late in autumn. The same rule applies to Rhipsalis, none of the species of which are happy when kept long dry. For the several species of Opuntia and Echinopsis, which are sufficiently hardy to be cultivated on a sunny rockery out of doors, it will be found a wise precaution to place either a pane of glass or a handlight over the plants in wet autumns and during winter, not so much to serve as protection from cold as to shield them from an excess of moisture at a time when it would prove injurious.

Temperature.—As the amount of heat required by the different species of Cactus varies very considerably, and as the difference between the summer and winter temperatures for them is often as great as it is important, it will be as well if we mention the temperature required by each when describing the species. It is true that the majority of Cactuses may be kept alive in one house where all would be subjected to the same temperature, but many of the plants would merely exist, and could not possibly flower. It would be easy to point to several instances of this unsatisfactory state of things. At Kew, for example, owing to the arrangements necessary for the public, it is found convenient to have the majority of the large collection of Cactuses in one house, where the plants present an imposing appearance, but where, as might be expected, a good number of the species very rarely produce flowers. The Cactuses which inhabit the plains of the Southern United States are subjected to a very high summer temperature, and a winter of intense cold; whilst on the other hand the species found in Central and South America do not undergo nearly so wide an extreme, the difference between the summer and winter temperatures of these countries being generally much less marked. A word will be said under each species as to whether it is tropical, temperate, or hardy, a tropical temperature for Cacti being in summer 70 degs., rising to 90 degs. with sun heat, night temperature 60 degs. to 70 degs., in winter 60 degs. to 65 degs. Temperate: in summer 60 degs., rising to 75 degs. with sun heat, night 60 degs. to 65 degs., in winter 50 degs. to 55 degs. The hardy species will, of course, bear the ordinary temperatures of this country; but, to enable them to withstand a very cold winter, they must be kept as dry as possible. In the colder parts of England it is not advisable to leave any of these plants outside during winter.

Insect Pests.—Notwithstanding the thickness of skin characteristic of almost every one of the Cactuses, they are frequently attacked by various kinds of garden pests when under cultivation, and more especially by mealy bug. There is, of course, no difficulty in removing such insects from the species with few or no spines upon their stems; but when the plants are thickly covered with clusters of spines and hairs, the insects are not easily got rid of. For Cactuses, as well as for other plants subject to this most troublesome insect, various kinds of insecticide have been recommended; but the best, cheapest, and most effectual with which we are acquainted is paraffin, its only drawback being the injury it does to the plants when applied carelessly, or when not sufficiently diluted. A wineglassful of the oil, added to a gallon of soft water, and about 2oz. of soft soap, the whole to be kept thoroughly mixed by frequently stirring it, forms a solution strong enough to destroy mealy bug. In applying this mixture, a syringe should be used, or, if the plants are to be dipped overhead, care must be taken to have the oil thoroughly diffused through the water, or the plant, when lifted out, will be covered with pure paraffin, which does not mix properly with water, but swims upon the surface if allowed to stand for a few moments. The plants should be laid on their sides to be syringed with the mixture, and after they have been thoroughly wetted, they may be allowed to stand for a few minutes before being syringed with pure water. Plants that are badly infested with mealy bug should be syringed with the paraffin mixture once a day, for about a week. It is easy to do serious harm to these plants by using a stronger solution than is here recommended, and also by not properly mixing the oil with the soap and water; and the amateur cannot, therefore, be too careful in his use of this excellent insecticide. It would be easy to recommend other insecticides, so called, for Cactuses; but whilst they are less dangerous to the plants, they are often as harmless as pure water to the insects.

For scale, which sometimes infests these plants, and which is sometimes found upon them when wild, the paraffin may be used with good effect.

Thrips attack Phyllocactus, Rhipsalis, and Epiphyllum, especially when the plants are grown in less shade, or in a higher temperature, than is good for them. Fumigation with tobacco, dipping in a strong solution of tobacco, or sponging with a mixture of soap and water, are either of them effectual when applied to plants infested with thrips. The same may be said of green-fly, which sometimes attacks the Epiphyllums.

A blight, something similar to mealy bug, now and again appears on the roots of some of the varieties of Echinocactus and Cereus. This may be destroyed by dipping the whole of the roots in the mixture recommended for the stems when infested by mealy bug, and afterwards allowing them to stand for a few minutes immersed in pure water. They may then be placed where they will dry quickly, and finally, in a day or two, repotted into new compost, first removing every particle of the old soil from the roots.

Diseases.—When wild and favourably situated as regards heat and moisture, the larger kinds of Cactus are said to live to a great age, some of the tree kinds, according to Humboldt, bearing about them signs of having existed several hundred years. The same remarkable longevity, most likely, is found in the smaller kinds when wild. Under artificial cultivation there are, however, many conditions more or less unfavourable to the health of plants, and, in the case of Cactuses, very large specimens, when imported from their native haunts to be placed in our glass houses, soon perish. At Kew, there have been, at various times, very fine specimens of some of the largest-growing ones, but they have never lived longer than a year or so, always gradually shrinking in size till, finally, owing to the absence of proper nourishment, and to other untoward conditions, they have broken down and rotted. This rotting of the tissue, or flesh, of these plants is the great enemy to their cultivation in England. When it appears, it should be carefully cut out with a sharp knife, and exposed to the influence of a perfectly dry atmosphere for a few days till the wound has dried, when the plant should be potted in a sandy compost and treated as for cuttings. Sometimes the decay begins in the side of the stem of the plant, in which case it should be cut away, and the wound exposed to a dry air. The cause of this decay at the base or in the side of the stems of Cactuses is no doubt debility, which is the result of the absence of some necessary condition when the plants are cultivated in houses or windows in this country.

Grafted plants, especially Epiphyllums, when worked on to Pereskia stocks, are apt to grow weak and flabby through the stem wearing out, or through the presence of mealy bug or insects in the crevices of the part where the stock and scion join, in which case it is best to prepare fresh stocks of Pereskia, and graft on to them the best of the pieces of Epiphyllum from the old, debilitated plant. It is no use trying to get such plants to recover, as, when once this disease or weakness begins, it cannot easily be stopped.

CHAPTER IV.

PROPAGATION.

ACTUSES may be multiplied from cuttings of the stems, from seeds, and also by means of grafting; this last method being adopted for those species which, under cultivation, are not easily kept in health when growing upon their own roots, or, as in the case of Epiphyllums, when it offers a means of speedily forming large and shapely specimens. From seeds the plants are generally freer in growth than when cuttings are used, although the seedlings are longer in growing into flowering specimens than large cuttings would be. To the amateur, the process of germination and development from the seedling to the mature stage, is full of interest and attraction, the changes from one form to another as the plant develops being very marked in most of the genera.

Seeds.—Good fresh seeds of Cactaceous plants germinate in from two to four weeks after sowing, if placed in a warm house or on a hotbed with a temperature of 80 degs. If sown in a lower temperature, the time they take to vegetate is longer; but, unless in a very low degree of heat, the seeds, if good, and if properly managed as regards soil and water, rarely fail to germinate. For all the kinds, pots or pans containing drainage to within 2 in. of the top, and then filled up with finely sifted loam and sand, three parts of the former to one of the latter, and pressed down moderately firm, will be found to answer. If the soil be moist at the time of sowing the seeds, it will not be necessary to water it for a day or two. The seeds should be scattered thinly over the surface of the soil, and then covered with about 1/8 in. of soil. Over this, a pane of glass may be placed, and should remain till the seedlings appear above the soil. Should the position where the seeds are to be raised be in a room window, this pane of glass will be found very useful in preventing the dry air of the room from absorbing all the moisture from the soil about the seeds. For the germination of Cactus, and indeed of all seeds, a certain amount of moisture must be constantly present in the soil; and after a seed has commenced to grow, to allow it to get dry is to run the risk of killing it.

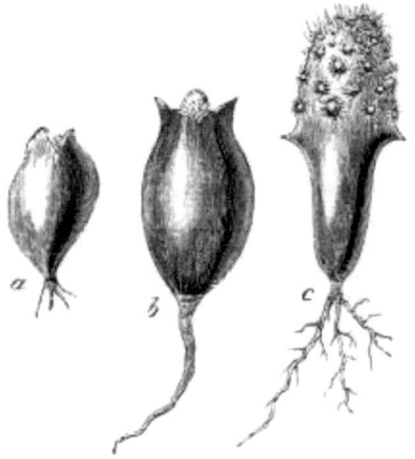

FIG. 4.—SEEDLINGS OF CEREUS.

a, One month after germination. *b*, Two months after germination. *c*, Three months after germination. (Magnified six times).

The seeds of Cactuses may be sown at anytime in the year; but it is best to sow in spring, as, after germinating, the young plants have the summer before them in which to attain sufficient strength to enable them to pass through the winter without suffering; whereas plants raised from autumn-sown seeds have often a poor chance of surviving through the winter, unless treated with great care. The seeds of all Cactuses are small, and therefore the seedlings are at first tiny globular masses of watery flesh, very different from what we find in the seedlings of ordinary garden plants. The form of the seedling of a species of Cereus is shown at Fig. 4, and its transition from a small globule-like mass of flesh to the spine-clothed stem, which characterises this genus, is also represented. At *a* we see the young plant after it has emerged from the seed, the outer shell of which was attached to one of the sides of the aperture at the top till about a week before the drawing was made. At *b*, the further swelling and opening out, as it were, of what, in botanical language, is known as the cotyledon stage of development, will be seen; a month afterwards, this will have assumed the shape of a very small Cereus. It is interesting to note how the soft fleshy mass which first grows out of the seed is nothing more than a little bag of food with a tiny growing point fixed in its top, and that, as the growing point increases, the food bag decreases, till finally the whole of the latter becomes absorbed into the young stem, which is now capable of obtaining nourishment by means of its newly-formed roots.

FIG. 5.—SEEDLINGS OF OPUNTIA, SHOWING MODE OF GERMINATION.

In the genus Opuntia, the cotyledon stage (see Fig. 5) of the plant is different from that of the Cereus, and is more like that of a cucumber. Still, though the form is different, the purpose of the two cotyledons and the juicy stem in the seedling Opuntia is the same as in the Cereus; and, as the growing point develops, the cotyledons shrivel up and fall off, the plant food they contained having passed into that part of the young seedling which was to be permanent. The seedlings of these two genera serve as an illustration of the process of germination from seed of all the Cactuses; and it must be evident that there is much that is singular and full of interest in raising these plants from seeds. As soon as the seedlings are large enough to be handled, they may be planted separately in small pots, using a compost similar to, but slightly coarser than, that in which the seeds were sown. The soil should be kept moist till the summer is over; and after that, till the return of warm sunny weather, it will be found safest to keep the seedlings on the dry side, a little water only to be given at intervals of a week, and only when the sun is shining upon the plants.

To obtain seeds from cultivated plants, it is necessary, in order to insure fertilisation that the top of the stigma (see Fig. 2) should be dusted over with the dust-like pollen from the anthers. This may be done by means of a small camel-hair brush, which should be moistened in the mouth and then pushed among the anthers till covered with pollen, which may then be gently rubbed on to the stigma. A warm, sunny morning is the most suitable time for this operation, as fertilisation takes place much more readily under the influence of bright sunshine than at any other time. Some of the kinds have their floral organs so arranged as to be capable of self-fertilisation; still, it is always as well to give them some assistance. The night-flowering species must, of course, be fertilised either at night or very early in the morning. By using the pollen from one kind for dusting on to the stigma of another, hybrids may be obtained, and it is owing to the readiness with which the plants of this family cross with each other, that so many hybrids and forms of the genera Epiphyllum and Phyllocactus have been raised. It would be useless to attempt such a cross as Epiphyllum with *Cereus giganteus,* because of their widely different natures; but such crosses as Epiphyllum with Phyllocactus, and *Cereus flagelliformis* with *C. speciosissimus,* have been brought about. To an enthusiast, the whole order offers a very good field for operations with a view to the production of new sorts, as the different kinds cross freely with each other, and the beautiful colours of the flowers would most likely combine so as to present some new and distinct varieties.

Cuttings.—No plants are more readily increased from stem-cuttings than Cactuses; for, be the cutting 20 ft. high, or only as large as a thimble, it strikes root readily if placed in a warm temperature and kept slightly moist. We have already seen how, even in the dry atmosphere of a museum, a stem of Cereus, instead of perishing, emitted roots and remained healthy for a considerable time, and it would be easy to add to this numerous other instances of the remarkable tenacity of life possessed by these plants. At Kew, it is the common practice, when the large-growing specimens get too tall for the house in which they are grown, to cut off the top of the stem to a length of 6 ft. or 8 ft., and plant it in a pot of soil to form a new plant. The old base is kept for stock, as it often happens that just below the point where the stem was severed, lateral buds are developed, and these, when grown into branches, are removed and used as cuttings. Large Opuntias are treated in the same way, with the almost invariable result that even the largest branches root freely, and are in no way injured by what appears to be exceedingly rough treatment. Large cuttings striking root so freely, it must follow that small cuttings will likewise soon form roots, and, so far as our experience—which consists of some years with a very large collection of Cactuses—goes, there is not one species in cultivation which may not be easily multiplied by means of cuttings. The nature of a Cactus stem is so very different from the stems of most other plants, that no comparison can be made between them in respect of their root-developing power; the rooting of a Cactus cutting being as certain as the rooting of a bulb. The very soft, fleshy stems of some of the kinds such as the Echinocactus, should be exposed to the air for a time, so that the cut at the base may dry before it is buried in the soil. If the base of a plant decays, all that is necessary is the removal of the decayed portion, exposure of the wound to the air for two or three days, and then the planting of the cutting in a dry, sandy soil, and placing it in a warm moist house till rooted. All cuttings of Cactuses may be treated in this way. If anything proves destructive to these cuttings, it is excessive moisture in the soil, which must always be carefully guarded against.

Grafting.—The object of grafting is generally either to effect certain changes in the nature of the scion, by uniting it with a stock of a character different from its own, which usually results in the better production of flowers, fruit, &c., or to multiply those plants which are not readily increased by the more ordinary methods of cuttings or seeds. In the case of Cactuses, however, we resort to grafting, not because of any difficulty in obtaining the kinds thus treated from either cuttings or seeds, as we have already seen that all the species of Cactuses grow freely from seed, or are easily raised from cuttings of their stems, nor yet to effect any change in the characters of the plants thus treated, but because some of the more delicate kinds, and especially the smaller ones, are apt to rot at the base during the damp, foggy weather of our winters; and, to prevent this, it is found a good and safe plan to graft them on to stocks formed of more robust kinds, or even on to plants of other genera, such as Cereus or Echinocactus. By this means, the delicate plants are raised above the soil whence the injury in winter usually arises, and they are also kept well supplied with food by the more robust and active nature of the roots of the plant upon which they are grafted. Grafting is also adopted for some of the Cactuses to add to the grotesqueness of their appearance; a spherical Echinocactus or Mamillaria being united to the columnar stem of another kind, so as to produce the appearance of a drum stick; or a large round-growing species grafted on to three such stems, which may then be likened to a globe supported upon three columns. As the species and genera unite freely with each other, it is possible to produce, by means of grafting, some very extraordinary-looking plants, and to a lover of the incongruous and "queer," these plants will afford much interest and amusement. Besides the above, we graft Epiphyllums, and the long drooping Cereuses, such as *C. flagelliformis*, because of their pendent habit, and which, therefore, are seen to better advantage when growing from the tall erect stem of some stouter kind, than if allowed to grow on their own roots. By growing a Pereskia on into a large plant, and then cutting it into any shape desired, we may, by grafting upon its spurs or branches a number of pieces of Epiphyllum, obtain large flowering specimens of various shapes in a comparatively short time. For general purposes, it is usual to graft Epiphyllums on to stems, about 1 ft. high, of *Pereskia aculeata;* pretty little standard plants being in this way formed in about a year from the time of grafting, As an instance of how easily some kinds may be grafted, we may note what was done with a large head of the Rat's-tail Cactus which had been grown for some years on the stem of *Cereus rostratus,* but which last year rotted off just below the point of union. On re-grafting this head on to the Cereus a little lower down, it failed to unite, and, attributing the failure to possible ill-health in the stock, we determined to transfer the Rat's-tail Cactus to a large stem of *Pereskia aculeata,* the result being a quick union and rapid, healthy growth since. Upon the same stock some grafts of Epiphyllum had previously been worked, so that it is probable these two aliens will form on their nurse-stem, the Pereskia, an attractive combination. In Fig. 6 we have a fine example of this kind of grafting. It represents a stem of *Pereskia Bleo* upon which the Rat's-tail Cactus and an Epiphyllum have been grafted.

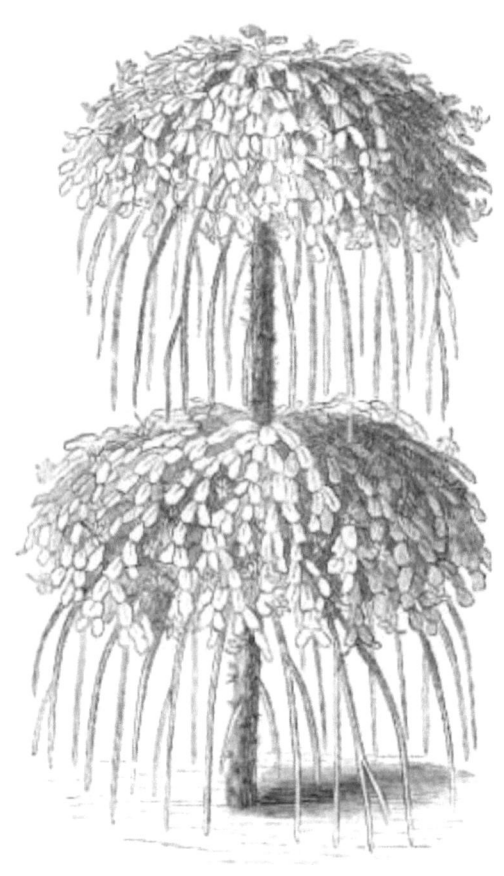

FIG. 6.—PERESKIA BLEO, WITH EPIPHYLLUM AND CEREUS FLAGELLIFORMIS GRAFTED UPON IT

For most plants the operation of grafting must be carefully and skilfully performed, but in the case of Cactuses very little skill is necessary if one or two rules, which apply to all kinds of grafting, are observed. The period of vigorous growth, and while the sap of both the stock and the scion is in motion, is the most favourable time for the operation. It is then only necessary, in order to bring about a speedy union, that the parts grafted should be cut so as to fit each other properly, and then bound or in some way fastened together so that they will remain in close contact with each other till a union is effected. A close atmosphere and, if possible, a little shade should be afforded the worked plants till the grafts have taken. The ligature used should not be bound round the graft too tightly, or it will prevent the flow of the sap; if bound tightly enough to hold the parts together and to prevent their slipping, that will be found quite sufficient.

Epiphyllums are treated as follows: Cuttings of Pereskia are rooted and grown on to the required size, and in the month of September they are headed down, the tops being used as cuttings. Grafts of Epiphyllum are then prepared by cutting them to the required length, usually about 6 in., and removing a thin slice of the fleshy stem on each side so as to form a flat wedge. The stem of Pereskia is then split down about 1 in. with a sharp knife, and into this the wedge of the graft is inserted, and fastened either by means of a small pin passed through the stem and graft about half-way up the slit, or by binding round them a little worsted or matting, the former being preferred. The worked plants are then placed in a close handlight or propagating frame, having a temperature of about 75 degs., where they are kept moist by sprinkling them daily with water; they must be shaded from bright sunlight. As soon as a union has been effected, which will be seen by the grafts beginning to grow, the ligature and pin should be removed, and the plants gradually hardened off by admitting air to the box, till finally they may be removed to the house where it is intended to grow them. In a cottage window this operation may be successfully performed if a box with a movable glass top, or a large bell glass, be used to keep the grafts close till they have taken.

For the spherical-stemmed kinds of Mamillaria, Cereus, Echinocactus, &c., a different method is found to answer. Instead of cutting the base of the graft to a wedge shape, it is simply cut across the base horizontally, or, in other words, a portion of the bottom of the graft is sliced off, and a stock procured which, when cut across the top, will about fit the wound at the base of the scion; the two sliced parts are placed together, and secured either by passing a piece of matting a few times over the top of the graft and under the pot containing the stock, or by placing three stakes around it in such a way that, when tied together at the top, they will hold the graft firmly in position. Another method is that of cutting the base of the scion in the form of a round wedge, and then scooping a hole out in the centre of the stock large enough to fit this wedge; the scion is pressed into this, and then secured in the manner above mentioned. To graft one spherical-stemmed kind on to three columnar-stemmed ones, the latter must first be established in one pot and, when ready for grafting, cut at the top into rounded wedges, three holes to correspond being cut into the scion. When fixed, the top should be securely fastened by tying it to the pot, or by means of stakes. For this last operation, a little patience and care are necessary to make the stocks and scions fit properly; but if the rules that apply to grafting are properly followed, there will be little fear of the operation failing. In the accompanying illustrations, we have a small Mamillaria stem grafted on to the apex of the tall quadrangular-stemmed, night-flowering Cereus (Fig. 7), and also a cylindrical-stemmed Opuntia worked on a branch of the flat, battledore-like Indian Fig (Fig. 8.)

FIG. 7.—GRAFT OF MAMILLARIA RECURVA ON CEREUS NYCTICALUS

FIG. 8.—GRAFT OF OPUNTIA DECIPIENS ON O. FICUS-INDICA

In the hands of a skilful cultivator, the different Cactuses may be made to unite with one another almost as easily as clay under the moulder's hands; whilst even to the amateur, Cactuses afford the easiest of subjects for observing the results of grafting.

CHAPTER V.

THE GENUS EPIPHYLLUM.

(From *epi* upon, and *phyllon*, a leaf).

T is now about a century since some of the most beautiful of Cactaceous plants came into cultivation in this country, and amongst them was the plant now known as *E. truncatum,* but then called *Cactus Epiphyllum;* the name Cactus being used in a generic sense, and not, as now, merely as a general term for the Natural Order. Introduced so early, and at once finding great favour as a curious and beautiful flowering plant, *E. truncatum* has been, and is still, extensively cultivated, and numerous varieties of it have, as a consequence, originated in English gardens. We do not use the seeds of these plants for their propagation, unless new varieties are desired, when we must begin by fertilising the flowers, and thus obtain seeds, which should be sown and grown on till the plants flower.

Epiphyllums have already "broken" from their original or wild characters, and are, therefore, likely to yield distinct varieties from the first sowing. In the forests which clothe the slopes of the Organ Mountains, in Brazil, the Epiphyllums are found in great abundance, growing upon the trunks and branches of large trees, and occasionally on the ground or upon rocks, up to an elevation of 6000 ft. It was here that Gardner, when travelling in South America, found *E. truncatum* growing in great luxuriance, and along with it the species known as *E. Russellianum,* which he sent to the Duke of

Bedford's garden, at Woburn, in 1839. These two species are the only ones now recognised by botanists, all the other cultivated kinds being either varieties of, or crosses raised from, them.

The character by which Epiphyllums are distinguished from other Cactuses, is their flattened, long, slender branches, which are formed of succulent, green, leaf-like branchlets, growing out of the ends of each other, to a length of from 3 ft. to 4 ft. As in the majority of Cactuses, the stems of Epiphyllum become woody and almost cylindrical with age, the axes of the branchlets swell out, and the edges either disappear or remain attached, like a pair of wings.

Cultivation.—Epiphyllums require the temperature of an intermediate house in winter, whilst, in summer, any position where they can be kept a little close and moist, and be shaded from bright sunshine, will suit them. Remembering that their habit, when wild, is to grow upon the trunks of trees, where they would be afforded considerable shade by the overhanging branches, we cannot be wrong in shading them from direct sunshine during summer. Some growers recommend placing these plants in a hot, dry house; but we have never seen good specimens cultivated under such conditions. All through the summer months, the plants should be syringed both morning and evening; but by the end of August they will have completed their growth, and should, therefore, be gradually exposed to sunshine and air.

It is advisable to discontinue the use of the syringe from September till the return of spring, but the plants should always be kept supplied with a little moisture at the root and in the air about them during the winter months. In this respect, these plants and the Rhipsalis are exceptions among Cactuses, as all the others are safest when kept dry during the cold, dull weather between September and April. The soil most suitable for them is a mixture of peat, loam, and sand, unless a light and fibrous loam be obtainable, which is, perhaps, the best of all soils for these plants, requiring only the addition of a little rotted manure or leaf-mould, silver sand, and some small brick rubble. The Pereskia stock is not a stout-rooted plant, and does not, therefore, require much root-room, although, by putting in plenty of broken crocks as drainage, the soil space in the pots may be reduced to what is considered sufficient for the plant. If small pots are used, the head of the plant is apt to overbalance the whole. The stems should be secured to stout stakes, and, if large, umbrella-like specimens are wanted, a frame should be made in the form of an umbrella, and the stem and branches fastened to it. Smaller plants may be kept in position by means of a single upright stake, which should be long enough to stand an inch or two above the head of the plant, so that the stoutest branches may be supported by attaching a piece of matting to them, and fastening it to the top of the stake. In the remarks upon grafting we mentioned the large pyramidal specimens of Epiphyllum which are grown by some cultivators for exhibition purposes; and, although these plants are much rarer at exhibitions now than they were a few years ago, yet they do sometimes appear, especially in the northern towns, such as Liverpool and Manchester.

It would not be easy to find a more beautiful object during winter than an Epiphyllum, 5 ft. or 6 ft. high, and nearly the same in width at the base, forming a dense pyramid of drooping, strap-like branches bearing several hundreds of their bright and delicate coloured blossoms all at one time, and lasting in beauty for several weeks. With a little skill and patience, plants of this size may be grown by any amateur who possesses a warm greenhouse; and, although it is not easy to manage such large plants in a room window, handsome little specimens of the same form may be grown if the window is favourably situated and the room kept warm in winter. Mr. J. Wallis, gardener to G.Tomline, Esq., of Ipswich, has become famous for the size and health of the specimens he has produced. Writing on the cultivation of Epiphyllums, Mr. Wallis gives the following details, which are especially valuable as coming from one of the most successful cultivators of these beautiful plants:

"The Epiphyllums here are grown for flowering in the conservatory, and are usually gay from the first week in November till February. During the remainder of the year, they occupy a three-quarter span-roof house, in which an intermediate temperature is maintained. All our Epiphyllums are grafted on the *Pereskia aculeata.* We graft a few at intervals of two or three years, so, if any of the older plants become sickly or shabby, they are thrown away, and the younger ones grown on. Some of the stocks are worked to form pyramids, and some to form standards. The height of the pyramids is 6 ft., and, to form these, six or eight scions are inserted. The heads of the standards are on stems ranging in height from 4= ft. down to 1= ft. To form these heads, only one scion is put on the stock. Some of our oldest pyramids are 4 ft. or 5 ft. through at the base, and the heads of the standards quite as much. When in flower, the heads of the latter droop almost to the pots. The pyramids occupy No.2 and No.4 sized pots, the standards 8's and 12's. Each plant is secured to a strong iron stake, with three prongs fitting the inside of the pot, and the Epiphyllum is kept well supported to the stake by ties of stout wire. After the plants are well established, they are easily managed, and go many years without repotting; but, of course, we top-dress them annually, previously removing as much of the old soil as will come away easily. We grow these plants with plenty of ventilation on all favourable occasions, and they are seldom shaded. During active growth, water is given freely, occasionally liquid manure; they are also syringed daily. After the season's growth is completed, water is given more sparingly, and syringing is dispensed with."

When grown on their own roots, Epiphyllums are useful for planting in wire baskets intended to hang near the glass; large and very handsome specimens form in a few years, if young rooted plants are placed rather thickly round the sides of the baskets, and grown in a warm house. Epiphyllums are employed with good effect for covering walls, which are first covered with peaty soil by means of wire netting, and then cuttings of the Epiphyllums are stuck in at intervals of about 1 ft. The effect of a wall of the drooping branches of these plants is attractive even when without their beautiful flowers; but when seen in winter, clothed with hundreds of sparkling blossoms, they present a most beautiful picture. Large plants of Pereskia may be trained over pillars in conservatories and afterwards grafted with Epiphyllums; in fact, there are many ways in which these plants may be effectively employed in gardens.

SPECIES.

E. truncatum (jagged); Bot. Mag. 2562.—Branchlets from 1 in. to 3 in. long, and 1 in. wide, with two or three distinct teeth along the edges, and a toothed or jagged apex (hence the specific name). The flowers are 3 in. long, curved above and below, not unlike the letter S; the petals and sepals reflexed, and exposing the numerous yellow anthers, through which the club-headed stigma protrudes; colour, a deep rose-red, the base of the petals slightly paler. The varieties differ in having colours which vary from almost pure white, with purplish tips, to a uniform rich purple, whilst such colours as salmon, rose, orange, and scarlet, are conspicuous among them.

FIG. 9.—EPIPHYLLUM RUSSELLIANUM

E. Russellianum (Russell's); Fig. 9.—This has smaller branchlets than the type plant *(E. truncatum)*, and is thus easily distinguished; they do not exceed 1 in. in length and = in. in width, whilst the edges are irregularly and faintly notched, not distinctly toothed, as in *E. truncatum*. The flowers are a little larger than in the older kind, and are not curved, whilst the petals are narrower; their colour is bright rosy-red. This species flowers rather later in the year than *E. trunca-*

tum, and may be had in blossom so late as the month of May or June. There are several varieties of it which have either larger and darker, or smaller and variously tinted flowers. Both the species will cross with each other, and probably many of the varieties enumerated by nurserymen have been obtained in this way.

VARIETIES.

The following is a selection of the best varieties, with a short description of the flowers of each:

E. bicolor (two-coloured).—Tube of flower white; petals purple, becoming almost white towards the base.

E. Bridgesii (Bridges').—Tube violet; petals dark purple.

E. coccineum (scarlet).—Bright scarlet, paler at the base of the petals.

E. cruentum (bloody).—Tube purplish-scarlet; petals bright scarlet.

E. Gaertneri (Gaertner's).—This is an interesting and beautiful hybrid, raised from Epiphyllum and a Cereus of some kind. The branchlets are exactly the same as those of *E. truncatum,* but the flowers are not like Epiphyllum at all, resembling rather those of Cereus or Phyllocactus. They are brilliant scarlet in colour, shaded with violet.

E. magnificum (magnificent).—Tube rosy-violet; petals dark red.

E. salmoneum (salmon-coloured).—Tube and base of petals white, rest salmon-red, shaded with purple.

E. spectabile (remarkable).—Tube and base of petals white; tips of petals carmine.

E. tricolor (three-coloured).—Tube salmon-red; petals red, centre purplish.

E. violaceum (violet).—Tube white; petals carmine, margined with violet-purple.

CHAPTER VI.

THE GENUS PHYLLOCACTUS.

(From *phyllon,* a leaf, and *Cactus*).

S in the case of the Epiphyllums, the principal character by which the Phyllocactus is distinguished is well described by the name, the difference between it and Epiphyllum being that in the former the flowers are produced along the margins of the flattened branches, whereas in the latter they are borne on the apices of the short, truncate divisions. If we compare any of the Phyllocactuses with *Cereus triangularis,* or with C. *speciosissimus,* we shall find that the flowers are precisely similar both in form and colour, and sometimes also in size.

In all the kinds the stem is compressed laterally, so as to look as if it had been hammered out flat; or sometimes it is three-angled, and the margins are deeply notched or serrated. These notches are really the divisions between one leaf and another, for the flat, fleshy portions or wings of the stems of these plants are simply modified leaves—not properly separated from each other and from the stem, but still to all intents and purposes leaves—which, as the plant increases and matures, gradually wither away, leaving the central or woody portion to assume the cylindrical stem which we find in all old Phyllocactuses. It is from these notches that the large, showy flowers are developed, just as in plants the flowers of which are borne from the axils of the leaves.

Under the names "Spleenwort-leaved Indian Figs," and "Winged Torch-thistles," as well as those here adopted, the most beautiful perhaps of all Cactuses, and certainly the most useful in a garden sense, have been cultivated in English gardens for more than 150 years; for it was in 1710 that the flowering of E. *Phyllanthus* was first recorded in English horticulture. Philip Miller grew it with many other Cactuses in the botanical garden at Chelsea which was founded by Sir Hans Sloane, in 1673, to be maintained "for the manifestation of the power, wisdom, and glory of God in the works of creation," and which still exists as the botanical emporium of the Apothecaries' Society. The majority of the gorgeous Phyllocactuses which we now possess are of only recent introduction, or are the result of cultivation and crossing.

The species are natives of various parts of tropical America, chiefly Mexico and Central America, where they are found generally growing, in company with Bromeliads and Orchids, upon the trunks of gigantic forest-trees. Phyllocactuses are therefore epiphytes when in a wild state, but under cultivation with us, they thrive best when planted in pots or in baskets—the latter method being adapted for one or two smaller kinds. It is easy to imagine the gorgeousness of a group of these plants when seen enveloping a large tree-trunk, clothing it, as it were, with balls of brilliant or pure white flowers. We are told by travellers of the splendours of a Cactus haunt during the flowering season, and those who have seen a well-managed pot specimen of Phyllocactus when covered with large, dazzling flowers, can form some idea of what wild plants are like when seen by hundreds together, and surrounded by the green foliage and festooning climbers which associate with them in the forests where they abound.

Cultivation.—For the following cultural notes we are indebted to a most successful grower of Cactuses in Germany, whose collection of Phyllocactuses is exceptionally rich and well managed: The growing season for these plants is from about the end of April, or after the flowers are over, till the end of August. As soon as growth commences, the plants should be repotted. A light, rich soil should be used, a mixture of loam, peat, and leaf-mould, or rotten manure with a little sand, being suitable. Small plants should have a fair shift; larger ones only into a size of pot which just admits of a thin layer of fresh soil. When pot-bound, the plants flower most freely, and it is not necessary to repot large specimens more often than about once every three years. When potted they should be placed in a sunny position in a close house or frame, and be kept freely watered. In bright weather they may be syringed overhead twice a day. For the first few days after repotting it is advisable to shade the plants from bright sunshine. A stove temperature is required until growth is finished. After this they should be gradually ripened by admitting more air and exposing to all the sunlight possible. During winter very little water is needed, just sufficient to prevent shrivelling being safest. Excess of moisture in winter is ruinous, as it often kills the roots, and sometimes causes the plant to rot off at the collar. The lowest temperature in winter should be 50 degs., lower than this being unsafe, whilst in mild weather it might be 5 degs. higher.

It is a bad plan to turn these plants round, in order, as some think, to ripen the growths properly. As a matter of fact, it does no good, but often does harm, by suddenly exposing the tender parts to the full force of sunlight.

The stems may be trained either in the form of a fan or as a bush. Old branches which have flowered and are shrivelling may be cut away in the spring.

Some fine specimens have been grown in pockets on old walls inside lean-to greenhouses, where the conditions have been favourable to the healthy growth and flowering of most of the species. When grown in this way, water must be supplied exactly as advised for plants grown in pots; if the pockets are not within easy reach of the watering pot, the plants can be watered by means of a heavy syringing.

Propagation.—For the propagation of the Phyllocactus either the whole plant may be divided at the base, or cuttings of the branches may be used; the latter, after having dried by remaining with their bases exposed to the air for a day or two, should be planted in small pots filled with very sandy soil; they may be placed on a dry, sunny shelf near the glass, and be slightly sprinkled overhead daily till rooted. Seeds, which sometimes ripen on cultivated plants, should be gathered as soon as the fleshy fruits have turned to a purplish colour, dried for a day or so, then sown in a light, porous soil, and placed in a warm frame or house to germinate.

SPECIES.

P. Akermanni (named after a Mr. Akermann, who introduced it from Mexico in 1829); Fig. 10.—Stem becoming cylindrical at an early age, and clothed with little clusters of spiny hairs; the branches are flattened out, and form broad, rather thin, blade-like growths, with the margins sinuately lobed (waved and notched). The flowers are large—over 6 in. in diameter—the petals, very acutely pointed and undulated along the edges; flower tube 2 in. long, with a few small scales scattered over its surface; stamens curved, clustered around the stigma, and almost hiding it. Colour of whole flower a rich scarlet, with a satin-like lustre. Flowers in June and July.

FIG. 10.—PHYLLOCACTUS AKERMANNI

This is one of the best-known kinds, having been extensively cultivated as an ornamental greenhouse plant till within the last few years. It was grown by several nurserymen for Covent Garden Market about eight years ago; small plants, about 1 ft. high, and bearing each from two to six flowers, finding much favour among the costermongers, as the plants could be bought at a low price, and, owing to their large, brilliant flowers, always sold well at a good profit. This species has been employed by the hybridists for the obtaining of new kinds, and some very handsome and distinct varieties have consequently been raised. As well as crossing with other species of Phyllocactus, *P. Akermanni* has been used in combination with several species of Cereus, good hybrids having been the result. As a compact-growing and free-flowering species, this may be specially recommended.

P. anguliger (angle-stemmed); Fig. 11.—The branches of this kind are distinguished by having the notches along their margins more like the teeth of a saw than the others. The habit is rather stiff and erect. The flowers are produced near the apex of the branches, and are composed of a curved tube 6 in. long, spreading out at the top to a width of 6 in., and surmounted by a whorl of pure white petals, in the centre of which are the stamens, rather few in number, and the large, ten-rayed stigma. The flowers are developed in December and January, and have a powerful and delicious odour. Introduced, in 1837, from West Mexico, where it is said to grow in oak forests.

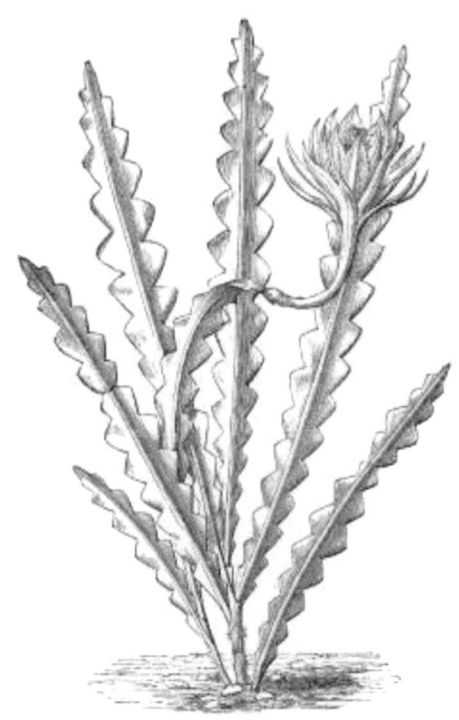

FIG. 11.—PHYLLOCACTUS ANGULIGER

P. (Disocactus) biformis (two-formed); Fig. 12.—This is a small plant, and is intermediate between this genus and the Epiphyllums. It possesses no particular beauty or distinctive character such as would render it of much value for garden purposes. The branches are short, rather narrow and drooping, the margins notched and tinged with red. The flowers are borne generally on the ends of the branches, and are drooping in habit; in form they are more like the Epiphyllums than the ordinary Phyllocactuses, as they have their petals arranged in a sort of tube about 3 in. long. The fruit is a red berry as large as a gooseberry. Honduras, 1839.

FIG. 12.—PHYLLOCACTUS BIFORMIS

P. crenatus (toothed); Bot. Reg. 3031.—A large-flowered and very beautiful species, rivalling, in the size and fragrance of its blossoms, the gigantic night-flowering *Cereus grandiflorus*. It grows to a height of about 2 ft., with round-based branches, the upper portion flattened out and the margins serrated. The flower tube is 4 in. long, brownish-green, as also are the sepals; petals 4 in. long, in a whorl, the points curving inwards; stamens and pistil erect, forming along with the petals a large star of a pale cream-colour. The beauty and fragrance of these flowers, which open in June, render them specially valuable for cutting and placing in rooms, where, notwithstanding their short duration, they never fail to win much admiration. Introduced from Honduras, in 1839. This fine species is one of the parents of the hybrids which have been raised both in this country and in America, where Mr. Hovey succeeded in obtaining some of the choicest as regards colour and size. Some of these latter were exhibited in London two or three years ago, and were much admired.

P. grandis (large-flowered). — The large, creamy-white flowers of this plant are like those of the night-flowering Cereus; and, in addition to the similarity in form and size between these two, there is a further one in the time when the flowers expand, this species, along with one or two others, opening its flowers after sunset; and although they remain in good condition till late on in the day following, and sometimes even longer, we may suppose that the proper flowering time is at night. The delicious almond scent of the flowers of this fine Cactus is so strong, that during the flowering period the atmosphere of the large Cactus-house at Kew Gardens is permeated with it, the large specimens there having usually a score or more flowers open together, the effect of which is truly grand. Even this number of flowers is, for this species, by no means extraordinary, specimens having been grown elsewhere, in pots only 8 in. across, with as many flowers open on each. From this it will be seen that *P. grandis* is one of the most useful kinds, its large, sweet-scented flowers, and its free-growing nature, rendering it of exceptional value as a decorative plant. Its branches are broad and notched along the margins, and the flowers are 1 ft. in length, including the tube, whilst across the broad, spreading petals they measure almost as much. Honduras. Introduced 1837 (?). Time of flowering, summer and autumn.

P. Hookeri (Hooker's); Bot. Mag. 2692, under *Cactus Phyllanthus*. — A robust-growing kind, often attaining to the size of a good shrub. Its flowers expand in the evening, and are sweet-scented. They are produced along the margins of the broad, flat, deeply-notched branches, the serratures being rounded instead of angled, as in some of the kinds. The tube of the flower is long and slender, no thicker than a goose quill, and covered with reddish scales; the petals are spreading, and form a cup 6 in. across; they are narrow, pointed, and pure white, the outer whorl, as well as the sepals, being tinged on the under side with a tawny colour. The stamens form a large cluster in the centre, and are bright yellow, the style being red and yellow. It is probable that this plant has been in cultivation for many years, as it was figured in the work quoted above under the name of one of the first introduced kinds of Phyllocactus, from which, however, it is abundantly distinct, as will be seen by a comparison of the descriptions of the two. There are, in the Kew collection, several large plants of *P. Hookeri* that flower annually during the summer and autumn. Brazil.

P. latifrons (broad-stemmed); Bot. Mag. 3813.—This is another large-growing species, as large at least as *P. Hookeri*, to which, indeed, it bears a close resemblance, both in flowers and in habit. Like that species, too, its date of introduction is not known, though it appears to have been cultivated in England at an early period. It may be grown so as to form a large shrub in a few years; or by cutting it back annually, or growing on young plants from cuttings every two years, nice little pot plants may be obtained; and as the plant produces flowers freely when in a small state, it is available for small greenhouses as well as for large ones. A fine specimen, such, for instance, as that at Kew, which is over 8 ft. in height, and well furnished with branches, is an attractive object when clothed with numerous creamy-white flowers, here and there tinged with red. The branches are from 4 in. to 5 in. broad, and deeply notched; the flowers are about 8 in. in length, and the same across the spreading petals. Mexico. Spring.

P. phyllanthus (leaf-flowering).—This species is now rarely seen in cultivation. As the oldest of the garden kinds it is, however, deserving of a little notice. Philip Miller grew it in his collection in 1710. The branches are broad and flat, the edges waved, not notched, and the flowers are composed of a thin tortuous tube, 9 in. in length, bearing at the top a whorl of recurved greenish petals, 1 in. long, with a cluster of whitish stamens and a green, club-shaped style and stigma. Brazil.

P. phyllanthoides (phyllanthus-like); Bot. Mag. 2092.—For the introduction of this handsome-flowered kind we are indebted to the great travellers and naturalists, Humboldt and Bonpland, who discovered it growing in the woods upon the trunks of old trees around Cartagena in South America. Plants of it were forwarded by them to France, where they flowered for the first time in 1811. From that time till now this species has been in favour as a garden plant, though it is, at the present time, much less common in English gardens than it deserves to be. The branches are broad, triangular when young, flat when old, about 1 ft. long by 2 in. wide, with shallow incisions, the serrations rather sharply angled. The height of the plant is from 2 ft. to 3 ft. The flowers are produced on the margins of the young branches, and are composed of a short, thick tube, not more than 2 in. in length, and short, dark, recurved scales; the petals are broad, pointed, and form a stellate cluster about 4 in. across; they are of a bright rose-colour, streaked with white, and shaded here and there with a darker colour of red. The stamens are numerous and pure white. The flowers open in the day-time, and are scentless; they last in perfection for two or three days, and may, therefore, be employed as cut flowers for vases, &c. Early summer.

HYBRIDS AND VARIETIES.

In addition to the cultivated species of Phyllocactus there are numerous hybrids and varieties, many of which are beautiful and distinct either in colour or in size of blossom.

The following is a selection of the best of them:

P. albus superbus (superb white).—The most beautiful of white-flowered kinds. Flowers fragrant, 6 in. across, resembling those of the night-blossoming *Cereus grandiflorus*; sepals greenish-white, petals pure white.

P. aurantiacus superbus (superb orange).—A compact plant, with numerous large, brick-red flowers, 5 in. to 6 in. in diameter.

P. Conway's Giant.—Flowers full, deep scarlet, about 8 in. in diameter.

P. Cooperi (Cooper's).—An English hybrid, remarkable for its large, beautiful yellow flowers.

P. Franzi (Franz's).—Flowers 3 in. to 4 in. across; petals numerous, outer ones scarlet, inner violet.

P. General Garibaldi.—Flowers very large, scarlet, tinged with orange on the reflex side.

P. grandiflorus (large-flowered).—Flowers bell-shaped, 4 in. across; sepals narrow, scarlet; petals incurved and of a fiery orange-scarlet colour.

P. Haagei (Haage's); Fig. 13.—Flowers about 5 in. across, flesh-coloured when first expanded, becoming carmine before fading.

FIG. 13.—PHYLLOCACTUS HAAGEI

P. ignescens (fiery).—Flowers 8 in. across, almost flat when expanded; petals numerous, deep brilliant scarlet.

P. Jenkinsoni (Jenkinson's).—Flowers medium in size, colour cherry-red.

P. Johnstonei (Johnstone's).—Flowers large, with broad scarlet petals.

P. Kaufmanni (Kaufmann's).—Flowers purplish-red, very large.

P. kermesina magnus (large scarlet).—An enormous-flowered kind, having produced blossoms which measured 10 in. across; petals vivid orange with a tip and central stripe of red; sepals blood-red.

P. Pfersdorffii. (Pfersdorff's).—Flowers as in *Cereus grandiflorus*, 8 in. to 10 in. across, very fragrant; petals white; sepals yellow, brownish outside.

P. Rempleri (Rempler's).—Branches three-angled; flowers with short, linear, incurved sepals; petals long, broad, arranged like a tube, colour salmon-red.

P. roseus grandiflorus (large rose-flowered); Fig. 14.—Flowers 6 in. long and broad, nodding, white.

FIG. 14.—PHYLLOCACTUS ROSEUS GRANDIFLORUS

P. Schlimii (Schlim's).—Branches three-angled; flowers large, sepals bright purple; petals broad, purple, tinged with scarlet.

P. splendens (splendid).—Flowers 8 in. across, purple-pink.

P. Wrayi (Wray's).—Flowers 5 in. long by 8 in. in diameter; sepals brown on the outside, yellow inside; petals yellowish-white, fragrant when first expanded.

CHAPTER VII.

THE GENUS CEREUS.

(From *cereus*, pliant; in reference to the stems of some species.)

VER 200 distinct species of Cereus are, according to botanists, distributed over the tropical and temperate regions of America and the West Indies, extending to the Galapagos, or "Tortoise" Islands, 200 miles off the coast of Peru. It was in these islands that the late Charles Darwin found several small kinds of Cereus, some of them growing near the snow-line in exposed situations on the highest mountains. In Mexico, *C. giganteus,* the most colossal of all Cacti, is found rearing its tall, straight, columnar stems to a height of 60 ft., and branching near the top, "like petrified giants stretching out their arms in speechless pain, whilst others stand like lonely sentinels keeping their dreary watch on the edge of precipices." In the West Indies most of the night-flowering kinds are common, their long, creeping stems clinging by means of aerial roots to rocks, or to the exposed trunks of trees, where their enormous, often fragrant, flowers are produced in great abundance, expanding only after the sun has set. Between these three distinct groups we find among the plants of this elegant genus great variety both in size and form of the stem and in the flower characters of the different species. A large proportion of the 200 kinds known are not cultivated in European gardens, and perhaps for many of them it is not possible for us to provide in our houses the peculiar conditions they require for their healthy existence. But there are a good many species of Cereus represented in gardens, even in this country, and among them we shall have no difficulty in finding many useful and beautiful kinds, such as may be cultivated with success in an ordinary greenhouse or stove. Lemaire, a French writer on Cactuses, groups a number of species under the generic name of Echinocereus; but as this name is not adopted in England, it is omitted here, all the kinds being included under Cereus.

THE NIGHT-FLOWERING SPECIES.

The most interesting group is that of the climbing night-flowering kinds, on account of their singular habit of expanding their flowers in the dark and of the very large size and brilliant colours of their flowers. In habit the plants of this set are trailers or climbers, their stems are either round or angled, and grow to a length of many feet, branching freely as they extend. By means of their roots, which are freely formed upon the stems, and which have the power of attaching themselves to stones or wood in the same way as ivy does, these kinds soon spread over and cover a large space; they are, therefore, useful for training over the back walls in lean-to houses, or for growing against rafters or pillars—in fact, in any position exposed to bright sunlight and where there is a good circulation of air. Soil does not appear to play an important part with these plants, as they will grow anywhere where there is a little brick rubble, gravel, or cinders for their basal roots to nestle in. They have been grown in the greatest luxuriance and have produced flowers in abundance with nothing more than their roots buried in the crumbling foundations of an old wall, upon which the stems were clinging. The chief consideration is drainage, as, unless the roots are kept clear of anything like stagnation, they soon perish through rot. During the summer, the stems should be syringed morning and evening on all bright days, whilst in winter little or no water will be required.

Like all other Cactuses, these plants may be propagated by means of large branches, which, if placed in a porous soil, will strike root in a few weeks. We saw a very large specimen of *C. triangularis*, which last autumn suddenly rotted at the base, from some cause or other, and to save the specimen, a mound was built up of brick rubble and soil, high enough to surround the base of the plant above the rotted part. In a few weeks there was a good crop of new roots formed, and the plant has since flowered most satisfactorily. With almost any other plant, this course would have proved futile; but Cactuses are singularly tenacious of life, the largest and oldest stems being capable of forming roots as freely and as quickly as the young ones.

C. extensus (long-stemmed); Bot. Mag. 4066.—This has long rope-like stems, bluntly triangular, less than 1 in. thick, with very short spines, arranged in pairs or threes, about 1 in. apart along the angles, and aerial roots. The flowers are developed all along the stems, and are composed of a thick, green, scale-clothed tube, about 3 in. long; the larger scales yellow and green, tipped with red, and a spreading cup formed of the long-pointed sepals and petals, the former yellow, green, and red, the latter white, tinted with rose. The flower is about 9 in. across. When in blossom, this plant equals in beauty the finest of the climbing Cactuses, but, unfortunately, it does not flower as freely as most of its kind. It is cultivated at Kew, where it has flowered once during the last five years. A native of Trinidad, whence it was introduced, and first flowered in August, 1843. Judging by the conditions under which it grows and blossoms in its native haunts, no doubt its shy-flowering nature under cultivation here is owing to the absence of a long continuance of bright sunshine and moisture, followed by one of drought and sunlight. If placed in a favourable condition as regards light, and carefully treated in respect of water, it ought to flower.

C. fulgidus (glittering); Bot. Mag. 5856.—In the brilliant deep scarlet of its large buds, and the bright orange-scarlet of the expanded flowers, this species stands quite alone among the night-flowering, scandent-stemmed Cereuses. Its one drawback is its shy-flowering nature, as it is rarely seen in blossom even when liberally treated, and along with the other kinds which flower so freely. The history of this plant is not known; but it is supposed to be a hybrid between *C. Pitajayi* or *variabilis* and one of the scarlet-flowered Phyllocactuses, or, possibly, *C. speciosissimus*. It first flowered at Kew, in July, 1870. Stems bright green, slow-growing, three or four-angled, about 2 in. wide; angles much compressed, so that a section of the stem shows a cross; margins notched, with clusters of short, hair-like spines at each notch. Flowers 6 in. long, and about the same across the top; tube covered with soft hairs and short deep-red scales, which are enlarged towards the top, where they spread out, and form, along with the petals, a large rosette of several whorls, arranged as in a semi-double rose, the centre being occupied by a brush-like cluster of greenish stamens, with the radiating stigma standing erect in the middle. It is to be regretted that the flowers are not more freely produced by cultivated plants.

C. grandiflorus (large-flowered); Bot. Mag. 3381.—There is scarcely any plant that makes a more magnificent appearance when in full blossom than this. A strong plant will produce many flowers together, but they do not remain long expanded, opening at seven or eight o'clock in the evening, and fading at sunrise the next morning; nor do they ever open again, even when cut and placed in warm water in a dark place. The closing of the flowers may, however, be retarded for a whole day by removing the bud before it is fully open and placing it in water. The stems are almost cylindrical, with four to seven slight ridges, or angles, which bear numerous tufts of wool and short stiff spines. Roots are thrown out from all parts of the stem, even when not in contact with anything. The flowers are developed on the sides of the stems, principally the younger, shorter ones; the flower tube is about 4 in. long by 1 in. in diameter, and is covered with short brown scales and whitish hairs; the calyx is 1 ft. across, and is composed of a large number of narrow sepals of a bright yellow colour inside, brown on the outside; the petals are broad, pure white, and arranged in a sort of cup inclosing the numerous yellow stamens and the club-shaped stigma. The flower has a delicious vanilla-like odour, which perfumes the air to a considerable distance. Flowers in July. Native of the West Indies. Introduced 1700, at which time it is said to have been cultivated in the Royal Gardens at Hampton Court.

C. Lemairii (Lemaire's); Bot. Mag. 4814.—In the size and fragrance of its blossoms, and also in the brilliancy of its colours, this species rivals *C. grandiflorus;* differing in the following particulars: the tube is covered with large green, crimson-edged scales instead of small brown scales and white hairs; the sepals do not spread out in a star-like manner, as in *C. grandiflorus,* and they are tinged with crimson; the stem of the plant shows a bluntly triangular section, and the angles are marked with a row of distant spines instead of the clusters of spines and wool in *C. grandiflorus*. In all other particulars, these two species are almost identical, so that where space is limited either the one or the other will be sufficient to represent both. *C. Lemairii* was introduced into England through Kew, whither a plant was sent in 1854 from the Royal Botanical Garden of Hanover, under the name of *C. rostratus.* It blossoms in the Kew collection every June, the flowers lasting for several hours after sunrise. Seeds are freely ripened by this plant. Native of Antigua (?)

C. Macdonaldiae (Mrs. MacDonald's); Bot. Mag. 4707.—A magnificent Cactus, producing flowers often 14 in. in diameter, with the same brilliant colours as are described under *C. Lemairii.* The stems are slender, cylindrical, not ridged or angled, bearing at irregular intervals rather fleshy tubercles instead of spines, and branching freely. Its flowers are produced on both young and old stems, several crops appearing in the course of the summer when the treatment is favourable. Roots are not so freely thrown out from the stems of this kind, and as the latter are slender and very pliant, they may be trained round a balloon trellis, so as to form handsome pot specimens, which, when in flower, may be carried into the house, where their large, beautiful flowers may be enjoyed. Writing of this species over thirty years ago, Sir Wm. Hooker said: "Certainly, of the many floral spectacles that have gratified lovers of horticulture at the Royal Gardens, Kew, of late years, few have been more striking than this to those who were privileged to see the blossoms in bud and fully expanded. The plant was received from Honduras through the favour of Mrs. MacDonald, and was planted at the back of the old Cactus-house, and trained against a wall. It first showed symptoms of blossoming in July, 1851. A casual observer might have passed the plant as an unusually large form of the 'night-blooming Cereus' *(C. grandiflorus),* but the slightest inspection of the stems and flowers, the latter 14 in. in diameter by 14 in. long, shows this to be a most distinct species."

C. Napoleonis (Napoleon's); Bot. Mag. 3458.—This is very like *C. grandiflorus*, and is slightly and not very agreeably perfumed. The flowers sometimes open very early in the morning and fade in the afternoon, so that they may be enjoyed during the day-time. The flower tube is 6 in. long, curved upwards, and clothed with rose-tinted scales, which become gradually larger towards the top, where they widen out into a whorl of greenish-yellow sepals, above which are the white petals forming a broad shallow cup, 8 in. across, with a cluster of yellow stamens in the centre. The stems are three-angled, light green, and bear clusters of short stiff spines along the angles at intervals of 2 in. Flowers in autumn. Mexico (?), 1835.

C. nycticalus (flowering at night); Fig. 15.—Stems four to six-angled, 2 in. wide, dark green, bearing little tufts of hair and thin white spines along the angles, and a profusion of aerial roots. Flowers as large as those of *C. grandiflorus*; tube covered with tufts of white hairs; sepals or outer whorl of segments bright orange, the inner pure white, and arranged like a cup. They open at about seven o'clock in the evening, and fade at seven on the following morning. This plant may still be met with in some old-fashioned gardens, but only rarely as compared with its popularity a generation ago, when it was to be found in almost every collection of stove plants. At that time, the flowering of this Cactus was looked upon as an event, and it was customary for the owner to invite his friends to meet and watch the development of the flowers, and enjoy to the full their almost over-powering but delicious fragrance. So bright are the colours of the flowers, that a sort of luminosity seems to surround them when at their best. Flowers in autumn. Mexico, 1834.

FIG. 15.—CEREUS NYCTICALUS

C. triangularis (three-angled); Bot. Mag. 1884.—This plant is easily recognised because of its stout triangular stems, which increase at a rapid rate and bear roots freely; by means of these roots they cling to almost any substance with which they come in contact. There are large examples of it in the Kew collection, where it bears numerous flowers annually, which open in the evening and close at about eight o'clock next morning. The flowers measure 1 ft. in length by about the same in width of cup, and are composed of a whorl of long narrow green sepals, with pale brown points, a cluster of pure white petals, bright yellow stamens, and a large club-like stigma;

they appear in autumn. Mexico. This species was cultivated at Hampton Court in 1690.

C. speciosissimus (most beautiful).—Although not a night-flowering kind, nor yet a climber, yet this species resembles in habit the above rather than the columnar-stemmed ones. It is certainly the species best adapted for cultivation in small greenhouses or in the windows of dwelling-houses, as it grows quickly, remains healthy under ordinary treatment, is dwarf in habit, and flowers freely — characters which, along with the vivid colours and large size of the blossoms, render it of exceptional value as a garden plant. Its stems are slender, and it may be grown satisfactorily when treated as a wall plant. For its cultivation, the treatment advised for Phyllocactuses will be found suitable. When well grown and flowered it surpasses in brilliancy of colours almost every other plant known. Specimens with thirty stems each 6 ft. high, and bearing from sixty to eighty buds and flowers upon them at one time, may be grown by anyone possessing a warm greenhouse. The stems are three to five angled, spiny, the tufts of spines set in little disks of whitish wool. The flowers are as large as tea saucers, with tubes about 4 in. long, the colour being an intense crimson or violet, so intense and bright as to dazzle the eyes when looked at in bright sunlight. When cut and placed in water they will last three or four days. April and May. Mexico, 1820. "Numberless varieties have been raised from this Cereus, as it seeds freely and crosses readily with other species. Many years ago, Mr. D. Beaton raised scores of seedlings from crosses between this and *C. flagelliformis,* and has stated that he never found a barren seedling. Much attention was given to these plants about fifty years ago, for Sir E. Antrobus is said to have exhibited specimens with from 200 to 300 flowers each. I have been informed that an extremely large plant of this Cereus, producing hundreds of flowers every season, is grown on the back wall of a vinery at the Grange, Barnet, the residence of Sir Charles Nicholson, Bart." (L. Castle).

THE SEMI-SCANDENT SPECIES.

These are characterised by a thin, drooping or trailing stem, and, though not strictly climbers, they may most fittingly be considered in a group by themselves. Some botanists have made a separate genus for them, viz., Cleistocactus, but for all practical purposes they may be grouped under the above heading, whilst popularly they are known as the Rat's-tail or Whipcord Cactuses. Two of them—viz., *C. flagelliformis* and *C. Mallisoni*—are generally grafted on the stem of some erect, slender Cereus or Pereskia, or they may be worked on to the stem of a climbing Cereus, such as *C. triangularis*, in such a way as to hang from the roof of a house. A large specimen of *C. flagelliformis*, growing from the climbing stem of *C. rostratus*, was, for a long time, conspicuous among the Cactuses at Kew, but owing to the decay of the "stock" plant, this fine specimen no longer exists. A large Pereskia, trained along the roof in the Cactus-house at Kew, has recently been grafted with a number of pieces of *C. flagelliformis*, which in a few years will, no doubt, form a handsome specimen. In the same establishment a specimen of *C. Mallisoni* is grafted on the stem of another kind, and is very attractive when in flower. *C. serpentinus* thrives well upon its own roots. For the cultivation of this little group, the instructions given for the climbing and other kinds may be followed.

C. flagelliformis (whip-formed).—Stems prostrate, or, when grafted on a tall stem, pendent, = in. in diameter, round, with numerous ridges almost hidden by the many clusters of fine bristle-like hairs. Flowers 2 in. long and 1 in. wide; colour bright rosy-red. In some parts of Germany this plant is one of the commonest of window ornaments, and it is so well grown by the peasants there, that the whole window space is completely screened by the numerous long, tail-like stems, 4 ft. or 6 ft. long, which hang from baskets. It is sometimes cultivated by cottagers in England, and we have seen a very fine specimen in a cottager's window in Gunnersbury. Without its pretty bright-coloured flowers, this Cactus has the charm of novelty in the form and habit of its stems, and as it is easily cultivated in a window through which the sun shines during most of the day, it is just the plant to grow for the double purpose of a screen and a curiosity. If planted in baskets, it should be potted in a porous loamy soil, and kept moist in the summer and perfectly dry in winter. Summer. Peru. Introduced 1690.

C. Mallisoni (Mallison's); Bot. Mag. 3822.—This is supposed to be of hybrid origin, a Mr. Mallison having sent it to Dr. Lindley to be named, and stating that he obtained it by fertilising flowers of *C. speciosissimus* with pollen from *C. flagelliformis*. Whatever its origin, it is a distinct kind, with stems similar to those of the last-named, but thicker and slightly less spiny, and flowers 4 in. long by 4 in. across the spreading petals, the whole being bright red with a cluster of pale yellow stamens protruding 1 in. beyond the throat. The flowers are produced from the sides of the stems, a few inches from the apex, and as they are borne in abundance and last three or four days each, a large specimen makes a very attractive display for several weeks in the summer. The plant at Kew, a large one, is grafted on the stem of *C. Macdonaldiae*, which is trained along a rafter, so that the stems of *C. Mallisoni* hang conspicuously from the roof.

C. serpentinus (serpent-like); Fig. 16.—When young, the stems of this plant are erect and stout enough to support themselves; but as they lengthen they fall over and grow along the ground, unless supported by a stake or wire; they have numerous ridges, with clusters of hair-like spines, which are usually purplish. Flowers large, handsome, fragrant; tube 6in. long, green; petals and sepals spreading and forming a star 3 in. in diameter, the petals purplish on the outside, and pinkish-white inside; stamens arranged in a sort of cup 1 in. deep. This plant rarely produces aerial roots. Small specimens are ornamental even when not in flower, the bright green, regularly ridged stem, with its numerous little clusters of fine spines, at the base of which are short tufts of a white woolly substance, being both curious and pretty. It flowers freely every summer. South America, 1814.

FIG. 16. – CEREUS SERPENTINUS

THE GLOBOSE AND COLUMNAR STEMMED SPECIES.

Many of these are unsuited for culture in ordinary plant-houses, whilst others are so rare that, although cultivated in botanical collections, they are not available for ordinary gardens, not being known in the trade. There are, however, a good many species that may be obtained from dealers in Cactuses, and to these we shall confine ourselves here. At Kew, the collection of Cereuses is large and diversified, some of the specimens being as tall as the house they are in will allow them to be, and the appearance they present is, to some eyes at least, a very attractive one. Such plants are: *C. candicans*, which is a cluster-stemmed kind, very thick and fleshy, and in shape like an Indian club; *C. chilensis*;—with tall hedgehog-skinned stems, the numerous ridges being thickly clothed with clusters of yellowish spines, which become dark brown with age; *C. Dyckii*, 10 ft. high, the stems thick and fleshy, with ridges 1= in. deep; *C. gemmatus*, a hexagonal, almost naked-stemmed species 10 ft. high; *C. strictus, C. peruvianus, C. geometrizans*, and *C. Jamacaru*, which are tall, weird-looking plants, 10 ft. or more high, some of them freely branched. The following is a selection of the largest-flowered and handsomest kinds:

C. Berlandieri (Berlandier's); Fig. 17.—A distinct and beautiful plant, of dwarf, creeping habit, forming a tuft of short branchlets springing from the main procumbent stems, none of which exceed 6 in. in length by > in. in thickness. They are almost round when old, the younger ones being slightly angled, and bearing, along the ridges, little tubercles, crowned with short spines. Even old stems are very soft and watery, and, on this account, it is necessary for the safety of the plant, in winter, that it should be kept absolutely dry. The flowers are produced on the young upright stems, and they are as much as 4 in. across. They are composed of a regular ring of strap-shaped, bright purple petals, springing from the erect bristly tube, and in the centre a disk-like cluster of rose-coloured stamens, the stigma standing well above them. In form the flowers are not unlike some of the Sunflowers or *Mutisia decurrens*. They are developed in summer, and on well-grown plants the display of blossom is exceptionally fine. This species is sometimes known as *C. repens* and *C. Deppii*. It is a native of South Texas and Mexico, where it is found growing in sandy or gravelly soils, on dry, sunny hill-sides. It should be grown in a cool greenhouse or frame, in a position where it would get plenty of sunshine to ripen its growth and induce it to flower. In winter it should be placed close to the glass, where the sun can shine full on it, and where it will be safe from frost. It will not thrive if wintered in a warm house. In April, it should be examined, repotted if the soil is sour, and kept watered as growth commences.

FIG. 17. — CEREUS BERLANDIERI

C. Blankii (Blank's); Fig. 18.—This is very similar to the *C. Berlandieri* in habit and stem characters, differing only in having longer, broader, less spreading petals, a club-shaped stigma, and in the colour, which is a deep rose, flushed in the throat with crimson. A comparison of the figures here given will show the differences better than any description. *C. Blankii* comes from Mexico at high elevations, and thrives under cultivation with the same treatment as the preceding. It is very common in Continental gardens, where it is grown out-of-doors, being protected from cold in winter by a handlight and straw. It flowers in summer.

FIG. 18.—CEREUS BLANKII

C. caerulescens (blue-stemmed); Bot. Mag. 3922.—An erect-growing, tall Cactus, rarely branching unless made to do so by cutting off the top of the stem; furrows and ridges about eight, the ridges prominent, waved, and bearing tufts of blackish wool, in which are set about a dozen black spines, = in. long; the stem when young and in good health is bluish in colour. Flowers springing from the ridges, about 8 in. long, the tube covered with reddish-grey scales, which pass upwards into the sepals; petals spreading, white, the margins toothed, and forming a spreading top, not unlike a large white single Camellia; the stamens are arranged in a sort of cup, and are yellow-anthered, with a large rayed yellow stigma in the middle. In the *Botanical Magazine* it is stated that the flowers of this species are equal and even superior to those of *C. grandiflorus*; but we have not seen flowers such as would bear out that statement. This species is too tall-stemmed to be recommended for windows or small greenhouses; but where room can be afforded it, the attractive colour of its stems, together with the size and beauty of its flowers, should win it favour. It blossoms in summer, generally about July, and is a native of Mexico. Introduced in 1841.

C. caespitosus (tufted); Fig. 19.—A dwarf species, the stem not more than 8 in. high by about 4 in. in diameter, sometimes branched, or bearing about its base a number of lateral growths, which ultimately form a cluster of stems—hence the name. The bark or skin of the stem is greyish-green, and the ribs, of which there are from a dozen to eighteen, are thickly covered with clusters of whitish wool and spines, the latter rose-tinted, and radiating in all directions. The flowers are produced on the top of the stems, and are short-tubed, the tube clothed with little bundles of spines; spread of the petals (from thirty to forty in each flower), 4 in.; colour deep rose; anthers and stigma forming an eye-like cluster, the former yellow, and the latter bright green. Flowered at Kew for the first time in 1882, but, although new to cultivation, it is becoming plentiful. Native of New Mexico and Texas. For windows or small greenhouses this is a most suitable plant, as it flowers freely and keeps in good health in an ordinary greenhouse temperature, always, however, requiring plenty of sunlight and rest during winter. By placing it upon a shelf near the glass from October to March, allowing it to remain perfectly dry, and afterwards watering it freely, the flowers should make their appearance early in summer. A plant with several stems, each bearing a large bright rose blossom, sometimes two, presents an attractive appearance.

FIG. 19.—CEREUS CAESPITOSUS

C. cirrhiferus (tendril-bearing).—A prostrate, branching-stemmed, small-growing kind, very proliferous, with roots along the main stems; branchlets upright, five-angled, with slightly raised points, or tubercles, upon which are ten short hair-like spines, arranged in a star, and surrounding three or four central erect spines, all whitish and transparent. Flowering branches erect, 4 in. high, by about 1 in. in diameter, bearing, near the apex, the large bright red flowers, nearly 4 in. in diameter, regular as a Sunflower, and lasting about a week. This species was introduced from Mexico in 1847. It is one of the best-known and handsomest of this group. It requires similar treatment to C. *Berlandieri*.

C. ctenoides (comb-like);

Fig. 20.—Stem 3 in. to 5 in. high, and about 3 in. in diameter, egg-shaped, unbranched, rarely producing offsets at the base. Ribs fifteen or sixteen, spiral, with closely-set cushions of stiff, whitish spines, which interlace and almost hide the stem; there are from fourteen to twenty-two spines to each cushion, and they are < in. long. Flowers produced on the ridges near the top of the stem; tube short, spiny; petals spreading, like a Convolvulus, 3 in. to 4 in. across, bright yellow; stamens yellow, pistil white. The flowers expand at about 9 a.m., and close again soon after noon. They are developed in June or July. This species is a native of Texas, and is rare in cultivation. When not in flower it might easily be mistaken for *Echinocactus pectinatus*. It should be grown in a sunny position, in a warm house or pit, all summer, and wintered on a shelf, near the glass, in a temperature of from 45 degs. to 50 degs. during winter. Under cultivation it is apt to rot suddenly at the base, more especially when old. Should this happen, the rotten parts must be cut away, and the wound exposed to the air in a dry house for a week or two.

FIG. 20.—CEREUS CTENOIDES

C. enneacanthus (eight-spined); Fig. 21.—Stem seldom more than 6 in. high by less than 2 in. in diameter, cylindrical in shape, bright green, simple when young, tufted in old specimens. Ribs shallow, broad, irregular on the top, with spine-cushions on the projecting parts; spines straight, yellowish-white, semi-transparent, variable in length, longest about 1 in. There are frequently as many as twelve spines in a tuft, although the specific name implies eight spines only. Flowers on the ridges near the top of the stem, with spiny tubes, spreading petals of a deep purple colour, and yellow stamens and pistil. They are developed freely in June and July. This is a soft-fleshed species, from Texas; it is not easily kept in health, and is therefore rarely seen. It should be treated as advised for *C. ctenoides*. Neither of these plants will flower unless it is grown in a sunny position as near to the roof-glass as is possible.

FIG. 21. — CEREUS ENNEACANTHUS

C. Fendleri (Fendler's). — One of the best of the dwarf-stemmed kinds. It has a pale green stem, about 6 in. high, rarely branching at the base, but often found growing in clusters. Ridges nine to twelve, running spirally round the stem, and bearing clusters of brown spines, some of them nearly 2 in. in length. Flowers composed of a tube 1 in. long, green, fleshy, and spiny, with a spreading cup-like arrangement of petals and sepals, 3 in. in diameter, and of a bright purple colour; stigma and anthers green. It produces its flowers in June. It was introduced from the mountainous region of New Mexico about five years ago, and has blossomed freely in several collections, notably in that of Mr. Loder, of Northampton, who has cultivated this and several other species from the same region in a sunny sheltered position out of doors, where, for several years, they have withstood winter's cold with no other protection than that afforded by an over-hanging wall. Mr. Loder says of *C. Fendleri* that it is the best of all Cactuses for cool treatment, as the flowers last more than a week, closing at night, and opening only in sunshine, when its rich purple colour is quite dazzling to the eye. It also blossoms freely under glass; but the colour of the flowers is not so vivid as when they are produced in full sunshine out of doors.

C. giganteus (gigantic); Fig. 22.—This is the most colossal of all Cactuses, in which respect it is chiefly interesting. Its stem, when young, is very similar to that of other dwarfer species, whilst, so far as is known, its flowers have not been produced under cultivation. It grows very slowly, a plant 6 in. high being eight or ten years old, so that, to attain its full development, a very long time indeed is necessary. When young, the stems are globose, afterwards becoming club-shaped or cylindrical. It flowers at the height of 10 ft. or 12 ft., but grows up to four or five times that height, when it develops lateral branches, which curve upwards, and present the appearance of immense candelabra. The flowers are 4 in. or 5 in. long, and about the same in diameter. There is a small specimen, about 3 ft. high, in the succulent collection at Kew. The appearance of a number of tall specimens of this wonderful Cactus, when seen towering high above the rocks and scrub with which it is associated, is described by travellers as being both weird and grand. Judging by the slowness of its growth, the prospect of seeing full-sized specimens of this species in English gardens is a very remote one, unless full-grown stems are imported, and this is hardly possible. Native of Mexico and California.

FIG. 22. — FLOWER OF CEREUS GIGANTEUS

C. Leeanus (Lee's); Bot. Mag. 4417. — A dwarf plant, the stems not more than 1 ft. in height, and about 5 in. in diameter at the base, tapering gradually towards the top, so that it forms a cone; the furrows number about a dozen, and the ridges are = in. high, the angles sharp, and clothed with clusters of pale brown spines, the central one 1 in. long, the others much shorter. The flowers are produced on the top of the stem, four or five together, and are large, handsome, brick-red in colour, the tube 2 in. long, clothed with yellowish, green-tipped scales, and little clusters of hair-like bristles. The arrangement of the petals, and the cluster of yellow anthers in the centre, give the flowers the appearance of Camellias, if looked at from above. Introduced from Mexico by Mr. Lee, of Hammersmith, in 1848, and flowered soon afterwards at Kew, in summer. Being a native of the higher, more northerly regions of Mexico, this species needs only to be protected from severe frosts; it has been known to bear a little frost without injury. For windows and greenhouses it is a very desirable plant.

C. leptacanthus (slender-spined); Fig. 23. — One of the most beautiful of all Cactuses, and one of the easiest to cultivate, the only drawback being that it rarely flowers under cultivation. In habit it is similar to C. *Berlandieri*. A plant 8 in. across bears about twenty short branches, each of which, under careful cultivation will produce several flowers in the months of May and June, and these, when expanded, last about eight days before withering; they close every afternoon, opening about ten o'clock in the morning. The petals are arranged in a single series, spreading so as to form a shallow cup, and are notched on the edges near the upper end. They are coloured a deep purple-lilac on the upper half, the lower part being white, like a large pied daisy. The stamens are pure white; the anthers orange-coloured, as also is the star-shaped stigma. The plant is a native of Mexico, and was introduced in 1860. It requires the same treatment as the preceding kinds. The illustration is sufficient to show the beauty of this little creeping Cactus, which, although so long known, is not grown in English gardens, though it is common enough in Continental collections.

FIG. 23.—CEREUS LEPTACANTHUS

C. multiplex (proliferous); Fig. 24.—A globose-stemmed species, becoming pear-shaped with age; height 6 in., by 4 in. in diameter; ridges angled, clothed with clusters of about a dozen spines, the central one longest. Flowers 6 in. to 8 in. long, and about the same across the spreading petals; tube clothed with small, hairy scales; the sepals long and pointed; petals 2 in. or more long, 1 in. wide, spreading out quite flat; stamens arranged in a ring, with the whitish-rayed stigma in the middle. This species flowers in autumn. It is a native of South Brazil, and was introduced in 1840. It thrives best when kept in a warm, sunny position in a window or heated greenhouse.

FIG. 24. – CEREUS MULTIPLEX

At Fig. 25 is a curious variety of the above, the stem being fasciated and divided into numerous crumpled, flattened branches. It is remarkable as a monster form of the type plant. So far as is known, neither this nor any other of the monster Cactuses produces flowers.

FIG. 25.—CEREUS MULTIPLEX CRISTATUS

C. paucispinus (few-spined); Bot. Mag. 6774.—A dwarf-stemmed species of recent introduction, and one which, owing to the beauty of its flowers and the hardy nature of the plant, is certain to find much favour among growers of Cactuses. The stem is about 9 in. high, by 2 in. to 4 in. in diameter, the base much wider than the apex, the ridges irregular, very thick and rounded, giving the stem a gouty or tumid appearance. Upon the prominent parts of these ridges are stellate tufts of long, pale brown spines, some of them nearly 2 in. long, and each tuft containing about eight spines. When young, the stems are more like some of the Mamillarias than the Cereuses. The flowers are developed near the top of the stem, two or three opening together; they are composed of a tube 2 in. long, clothed with long spines and large, green, scaly sepals below, the latter gradually enlarging till at the top they become as large as the petals, which are 2 in. long, with a spread of nearly 3 in., rounded at the tips, and coloured deep blood-red, tinged with orange inside. The stamens are clustered together sheaf-like, with the dark green stigmas protruding through them. This is a native of New Mexico, whence it was introduced in 1883, and flowered in May. Mr. Loder, of Northampton, has successfully cultivated it in a cool frame in the open air, and it has also grown well in the Kew collection when treated in a similar way. This suggests its hardiness and fitness for window cultivation. Owing to the watery nature of the stems, it is necessary that they should be kept quite dry during the winter.

C. pentalophus (five-winged); Bot. Mag. 3651.—As the name denotes, the stem of this erect-growing, somewhat slender species has five very prominent sharp-edged ridges, along which are little clusters of small spines about = in. apart; the stem is 1 in. in diameter, and the angles are wavy. The flowers are about 3 in. wide, spreading, the petals, broad and overlapping, rose-coloured, except in the centre of the flower, where they become almost pure white; the anthers are yellow, whilst the colour of the rayed stigma is purplish-blue. A native of Mexico, introduced and flowered in 1838. For its cultivation, the temperature of a warm greenhouse is required, though during summer it may be placed in a sunny position in a frame out of doors. If grown in windows, it should be kept through the winter in a room where there is a fire constantly.

C. peruvianus (Peruvian).—A tall-grower, the stems fleshy when young, and very spiny. The ridges on the stem number from five to eight, with stellate bundles, about 1 in. apart, of small, stiff black spines. The flowers appear upon the upper portion of the stem, and are 5 in. across, the petals pure white above, tinged with red below, and forming a large saucer, in the middle of which the numerous stamens, with yellow anthers, are arranged in a crown. There is something incongruous in the tall, spine-clothed, pole-like stem, upon which large, beautiful, water-lily-like flowers are developed, looking quite out of place on such a plant. Flowers in spring and early summer. It requires warm greenhouse or stove treatment. There are some fine examples of this species at Kew. A variety of this species, with a fasciated or monstrous habit, is sometimes cultivated. Introduced in 1830.

C. pleiogonus (twisted-angled); Fig. 26.—An erect cylindrical-stemmed species, from 6 in. to 1 ft. high by 4 in. in diameter, with from ten to fourteen angles or ridges; these are somewhat tumid, and marked with depressions, from which the star-like clusters of spines spring, about a dozen spines in each cluster, the central one much the largest. The flowers are about 8 in. long, the tube being rather thick and cylinder-like, expanding at the top, so as to form a sort of cup, in which the petals are arranged in several rows, with the middle filled by the numerous stamens, surmounted by the club-like pistil. The colour of the flowers is purple-red. This species appears to have first found its way into cultivation through some Continental garden, its native country being unknown. It thrives only in a warm house, developing its flowers in summer.

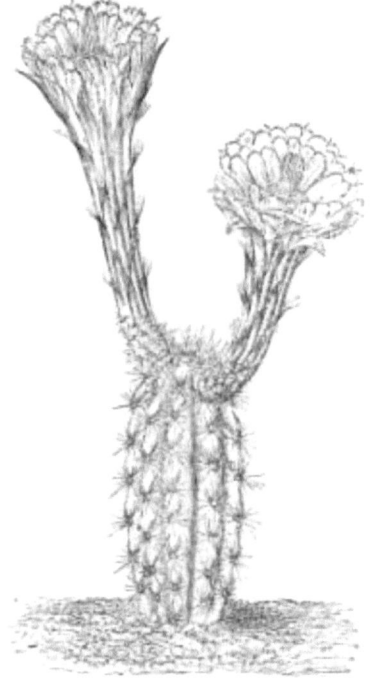

FIG. 26.—CEREUS PLEIOGONUS

C. polyacanthus (many-spined).—A newly-introduced species, from El Paso, in Mexico, where it is common on the sand ridges and stony hills. Stem 10 in. high, 2 in. to 4 in. wide, pale green or glaucous, with about eight ridges, the spines being placed along the angles in clusters of half a dozen or so, and about = in. apart. The flowers are 2 in. to 3 in. long; the tube spiny; the petals semi-erect and concave, rounded at the tip, and forming a shallow cup or wine-glass-like flower; the colour of the petals is deep blood-red. This beautiful Cactus is exceptional in the length of time its flowers remain expanded and fresh, lasting a week or more; and as the plant is very free flowering, there is usually a beautiful display of rich red blossoms for about six weeks. It may be grown in a cool greenhouse or window, requiring no artificial heat beyond what would be necessary to insure its protection from frost. It flowers in spring.

C. procumbens (trailing); Fig. 27.—This is a very pretty little Cactus, with spreading prostrate stems, from which upright branches grow to a height of 3 in. or 4 in.; they are = in. thick, generally only four-angled or square, with small spines in tufts along the angles. The flowers are developed on the ends of the branches, and are 3 in. long and wide, the sepals spreading and recurved, as in a Paris daisy, their colour being bright rose purple. The anthers form a corona-like ring, inclosing the upright, rayed stigma. A native of Mexico; flowers in May and June. In its native haunts we learn that this little Cactus is very free-flowering, but under cultivation in this country it flowers only rarely. It thrives best when grown in a dry, sunny greenhouse, and kept perfectly dry during autumn and winter. If allowed to get wet in that season, it is apt to rot, the stems being soft and watery.

FIG. 27.—CEREUS PROCUMBENS

C. reductus (dingy); Bot. Mag. 4443.—Stem erect, sometimes 3 ft. high, and about 4 in. wide, deeply furrowed, the furrows usually numbering about fourteen; the ridges tumid and irregular, and coloured a dingy glaucous-green. Spines embedded in a tuft of grey wool, about a dozen spines in each cluster, 1 in. long, a few of them only half that length. Flowers on the top of the stem, three or four opening together, each being 3 in. long and wide; the tube short and scaly, with overlapping sepals and saw-edged petals, which are white, slightly tinged with rose. Stamens filling the whole of the flower-cup, bright yellow. A native of Mexico, introduced in 1796, flowering in summer. This species was evidently a favourite many years ago, but it is rare with us now. It thrives in a house where the winter temperature does not fall below 45 deg., requiring no water at that time, but a liberal supply in the summer when growth is being made, and all the sunlight possible. When without its star-shaped, handsome flowers, the stem is remarkably ferocious-looking, the spines upon it being quite as thick and as strong as on a hedgehog.

C. repandus (undulated); Fig. 28.—Stem erect, 10 ft. or more high, unbranched, unless compelled to do so by the removal of the top. Ribs eight or nine in number, rounded, somewhat undulated, and bearing spine-tufts nearly 1 in. apart; each tuft contains about ten spines, which are almost equal in length, fine, stiff, brown, and persistent; there is a little cushion of white wool about the base of the spines. Flowers produced on the side, within a few inches of the top of the stem; they are composed of a scaly tube, 4 in. long, a circular row of spreading, incurved, pale brown sepals, and two rows of broad, overlapping, snow-white petals; stamens white, with yellow anthers; stigma yellow. The flowers, developed in summer, are very beautiful, but, unfortunately, each lasts only a few hours. A native of the West Indies, and an old introduction to English gardens (1720), but rare in cultivation now. It requires the treatment of a stove all the year round.

FIG. 28.—CEREUS REPANDUS

C. Royeni (Royen's); Bot. Mag. 3125.—This plant is not one of the handsomest as regards flowers; but its stems are ornamental, and the form of the flowers is such as would please those who admire the curious. The stem is erect, several feet high, 2 in. in diameter, with about ten acute ridges, along which are little tufts of white wool about the base of the clustering spines, which are dark brown and 1 in. long. The flower-tube is 2 in. long, thick, spineless, scaly, the scales becoming large near the top of the flower, where they form a cup-like whorl, enclosing the small rose-coloured petals, the stamens being white. Introduced from New Grenada, in 1832. It flowers in spring and summer. It should be grown in a stove.

C. variabilis (variable); Bot. Mag. 4084, under the name of *C. pitajaya*. — A tall-growing plant, rather straggling in habit, branching freely, the stems usually four-winged, but sometimes with three, five, or more, constricted at intervals, as in Phyllocactus, the wings spiny along the edges; spines 1 in. long. Flowers on the sides of the stems, rather low down, long-tubed; large, showy; tube 6 in. long, smooth, fleshy, with a few scales near the top, and a whorl of greenish, strap-shaped, pointed sepals, the petals spreading, with toothed margins and a long acute point, white or cream-coloured; anthers yellow. A native of various parts of South America and the West Indies, but always close to the sea. It flowers in July; the flowers, which open generally in the evening, remain expanded all night, and close before noon the day following. This species requires tropical or warm house treatment. There are some old plants of it in the Kew collection, where it flowers annually. Except for large houses, this species is not recommended for general cultivation, as it blossoms only after attaining a good size, and the stems, when old, are not at all ornamental.

CHAPTER VIII.

THE GENUS ECHINOCACTUS.

(From *echinos*, a hedgehog, and *Cactus*.)

ANY of the plants included in the genus Echinocactus are very similar in habit and stem-characters to the Cereus. Botanists find characters in the seed vessel (ovary) and in the seeds by which the two genera are supposed to be easily separable; but, so far as can be made out by a comparison of their more conspicuous characters, there is very little indeed to enable one to distinguish the two genera from each other when not in flower. A comparison of the figures given in these pages will show that such is the case.

The name Echinocactus was given to *E. tenuispinus*, which was first introduced into English gardens in 1825. The spiny character of this species is surpassed by that of many of the more recently introduced kinds; still it is sufficient to justify its being compared to a hedgehog. Some of the kinds have spines 4 in. long, broad at the base, and hooked towards the point, the hooks being wonderfully strong, whilst in others the spines are long and needle-like, or short and fine as the prickles on a thistle. The stems vary much in size and form, being globose, or compressed, or ovate, a few only being cylindrical, and attaining a height of from 5 ft. to 10 ft. They are almost always simple—that is, without branches, unless they are compelled to form such by cutting out or injuring the top of the stem; the ridges vary in number from about five to ten times that number, and they are in some species very firm and prominent, in others reduced to mere undulations, whilst in a few, they are separated into numerous little tubercles or mammae. The species are nearly all possessed of spines, which are collected in bundles along the ridges of the stem. Generally, the flowers are about as long as wide, and the ovary is covered with scales or modified sepals. The fruit is succulent, or sometimes dry, and, when ripe, is covered with the persistent calyx scales, often surrounded with wool, and usually bearing upon the top the remains of the withered flower. The position of the flowers is on the young part of the stem, usually being perched in the centre, never on the old part, as in some of the Cereuses. The flowers open only under the influence of bright sunlight, generally closing soon after it leaves them.

The geographical distribution of the species, of which over 200 have been described, extends from Texas and California to Peru and Brazil; they are in greatest abundance in Mexico, whence most of the garden kinds have been introduced. The conditions under which they grow naturally vary considerably in regard to temperature and soil; but they are all found in greatest numbers and most robust health where the soil is gravelly or sandy, and even where there is no proper soil at all, the roots finding nourishment in the clefts or crevices of the rocks. As a rule, the temperature in the lands where they are native is very high during summer, and falls to the other extreme in winter, some of the species being found even where frost and snow are frequent; the majority of them, however, require what we would call stove treatment.

Turning now to a consideration of those kinds known as garden plants, we find that comparatively few of the species known to botanists are represented in English collections, though, perhaps, we may safely say that not one of the kinds known would be considered unworthy of cultivation except by those who despise Cactuses of whatever kind. Their flowers are conspicuous both in size and brilliancy of colour; and in the curious, grotesque, and even beautifully symmetrical shapes of their stems, one finds attractions of no ordinary kind. The stem of *E. Visnaga* shown at Fig. 48 may be taken as an instance of this—apart from the cluster of star-like, bright yellow flowers seen nestling upon the top of their spine-protected dwelling, the whole suggesting a nest of young birds. This plant is indeed one of the most remarkable of the Echinocactuses, owing to the size and number of its spines—which are 3 in. long, almost as firm as steel, and are used by the Mexicans as toothpicks—and to the gigantic size and great weight of the stem. The following account of a large specimen of this species introduced to Kew in 1845, is taken from an article from the pen of the late Sir Wm. Hooker in the *Gardeners' Chronicle* of that year. This gigantic plant was presented to the nation, in other words to Kew, by F. Staines, Esq., of San Luis Potosi. Such was its striking appearance, that it was stated that, if exhibited in the Egyptian Hall, Piccadilly, some hundreds of pounds might be realised by it. In a letter from Mr. Staines, here quoted, our readers will perceive how difficult it often is to obtain living specimens of these plants from their native habitats. He writes: "I mean to have a large specimen of *E. Visnaga* deposited in a strong box, sending the box first to the mountain where the monsters grow, and placing it on the springs of a carriage which I shall despatch for that purpose. My monstrous friend cannot travel any other way, from his stupendous size and immense ponderosity, which cannot be adequately calculated for here, where the largest machine for conveying weights does not exceed sixteen arrobes, or 400lb. This enormous plant will require twenty men at least to place it upon the vehicle, with the aid of such levers as our Indians can invent. It grows in the deep ravines of our loftiest mountains, amongst huge stones; the finest plants are inaccessible to wheeled vehicles, and even on horseback it is difficult to reach them. I shall pack him carefully in mats before applying to his roots the crowbars destined to wrench him from his resting place of unknown centuries. He will have to travel 300 leagues before he reaches Vera Cruz." Being too large to be packed in a box, it was first surrounded with a dense clothing of the Old Man's Beard or Spanish moss (*Tillandsia usneoides*)—and a better covering could not have been devised—and well corded. Fifteen mats, each as large and as thick as an ordinary doormat, formed the exterior envelope. When unpacked on its arrival at Kew, this monster Cactus was seen as perfect, as green, and as uninjured as if it had been that morning removed from its native rocks, its long, rope-like roots arranged in coils like the cable of a ship. When placed in scales it weighed 713lb., its circumference at 1 ft. from the ground was 4= ft., and its total height, 8 ft. 7 in.; the number of ridges was forty-four, and on each ridge were fifty bundles of spines, four spines to each bundle.

Thus there were 8800 spines or toothpicks, enough for the supply of an army. A still larger specimen was a year or so later successfully brought to Kew, and which weighed 1 ton; but this, as well as the smaller one, survived only a short time. There have been numerous other large specimens of this Cactus in English gardens lately, all of them, however, succumbing to the unfavourable conditions of our climate. Mr. Peacock, of Hammersmith, recently possessed two large plants of E. *Visnaga,* one of which weighed nearly 5cwt., and measured 8 ft. 6 in. in circumference.

Cultivation.—The soil for Echinocactuses should be similar to that recommended for the Cereuses, as also should be the treatment as regards sunlight and rest. It cannot be too clearly understood that during the period between October and March these plants should be kept perfectly dry at the root, and in a dry house, where the temperature would not fall below 50 deg. There is no occasion for re-potting the Echinocactuses every year, it being by far the safest plan to allow them to remain in the same pots several years, should the soil be fresh and the drainage perfect.

All the larger-stemmed kinds may be kept in health when grown on their own roots; but for some of the smaller species it is a good plan to graft them upon the stem of some of the Cereuses, *C. tortuosus* or *C. colubrinus* being recommended for the smaller kinds, and for the larger *C. peruvianus*, *C. gemmatus*, or any one the stem of which is robust, and of the right dimensions to bear the species of Echinocactus intended to be grafted. Some growers prefer to graft all the small Echinocactuses upon other kinds, find certainly, when properly grafted, they are safer thus treated than if grown on their own roots. In grafting, the two stems (stock and scion) must be cut so that their edges meet, and in securing them two or three stakes must be placed in such a way as to afford support to the graft and hold it firmly in position.

Propagation.—Besides grafting, cuttings of the stems may be utilised for the multiplication of Echinocactuses, first removing the upper portion of the stem and putting it into soil to root, and afterwards, as lateral stems develop on the old stock, they may be cut away with a sharp knife, and treated in a similar manner. Should a plant become sickly, and look shrivelled and cankered at the base, it is always best to cut away the healthy part of the stem, and induce it to form fresh roots, thus giving it a new lease of life. Seeds of these plants may be obtained from dealers, more especially Continental nurserymen, and to watch the gradual development of the plant from the seedling is both interesting and instructive. The seeds should be sown in soil, and kept moist and warm; in about a month after sowing, the little pea-like, green balls will be seen pushing their way through the thin covering of soil, and gradually but slowly increasing in size, their spines also increasing in number and strength, the ridges forming according to the character of the species, till, finally, they assume the mature characters of the plant, both in stem and habit. The flowers, of course, appear according to the length of time it takes for the species to grow to flowering size.

SPECIES.

E. brevihamatus (short-hooked).—Several kinds of Echinocactus are distinguished from the rest in having the ridges divided into tubercles, which are often globular and arranged in a spiral round the stem, as in the genus Mamillaria; to this section the present species belongs. The stem is almost sphere-shaped, from 4 in. to 6 in. high, the tuberculated ridges about < in. deep, and upon each tubercle is a tuft of about a dozen brown, radiating spines, with a long central one hooked at the point. The flowers are borne in clusters on the top of the stem, three or four opening together; they are 1 in. in length, and the same across the spreading petals, which are pink, shaded with deep rose. A native of the mountainous regions of South Brazil; introduced about 1850. Flowers in summer. This pretty little plant will thrive if placed upon a shelf in a greenhouse where it will have full sunshine during the greater part of the day. It grows very slowly, especially when on its own roots, but succeeds better when grafted on another kind.

E. centeterius (many-spined); Bot. Mag. 3974.—This has a conical-shaped stem, 6 in. high by 4 in. wide, with about fourteen ridges, which are notched, and bear star-shaped clusters of pale brown spines, = in. long. The blossoms are borne rather thickly on the summit of the stem, from six to nine flowers being sometimes open together; and as they are each nearly 3 in. across, and of good substance, they present an attractive appearance. The petals are of a deep straw-colour, with a reddish streak down the centre, and 1= in.long, with the apex notched or toothed. The stamens are spirally coiled round the stigma, which is club-shaped and white. This species is probably a native of Mexico, and was first flowered in England at Kew, in 1841. A cool, dry greenhouse suits it best; or it may be grown in a sunny room window where frost would not be allowed to reach it in winter. Unless subjected to very dry treatment during the winter months, and also kept in a position where all the sunlight possible would reach it—even when at rest—there is not much chance of this plant producing its large flowers. It may be kept alive by giving it uniform treatment all the year round, but it would never flower.

E. cinnabarinus (cinnabar-flowered); Bot. Mag. 4326.—This is another of the Mamillaria-like kinds, and is remarkable for the depressed form of its stem, which may be likened to a sea urchin, both in size and shape. Old plants are from 6 in. to 8 in. in diameter, and about 4 in. high; the spiral formed by the tubercles rises very gradually, and each of the latter is surmounted by a tuft of strong, brown, radiating spines, imbedded in a little cushion of wool. The flowers spring from the outside of the depressed top of the stem, two or three opening together and forming a beautiful picture, both as to size and colour. The tube is short and green, with a row of long green sepals at the top, and above these the petals, which are 2 in. long, overlapping, recurved, the edges toothed, and the colour a brilliant cinnabar-red. The stamens are in two series, very numerous, and the anthers are bright yellow. Looking at the flattened, spiny stem, it seems impossible that such large, handsome flowers should be produced by it. A native of Bolivia; introduced about 1846. It blossoms in July, and may be grown on a shelf in a cool greenhouse, as advised for the *E. centeterius*.

E. concinnus (neat); Fig. 29.—A small species with a globose stem, 2 in. high and 3 in. wide, and about twenty ridges, which are rounded, rather broad, each bearing about half-a-dozen little bunches of spines arranged in a star. The flowers are numerous, as large as, or larger than, the stem, being 3 in. long and broad, the tube covered with brown hair-like spines, and having a few reddish scales, whilst the petals are in several rows, overlapping, with pointed tips, and are coloured dark yellow with a red streak down the centre. Several flowers are sometimes developed together on a little stem, when they have the appearance of being much too large for so small a plant to support. The pale green of the stem and its brown spines contrast prettily with the handsome yellow flowers, which are brightened by the streaks of red on the petals and the clear red colour of the stigma. It is a native of Mexico, and was introduced about 1840, flowering early in summer. It requires a warm greenhouse temperature all the year round, with, of course, plenty of sunshine. It may be grafted on the stem of an erect-growing Cereus, such as *C. serpentinus* or *C. Napoleonis*, the stock to be not higher than 6 in., and about as wide as the plant of *E. concinnus* is at the base.

FIG. 29. — ECHINOCACTUS CONCINNUS

E. coptonogonus (wavy-ribbed); Fig. 30. — Stem globose, seldom more than 5 in. in diameter, depressed on the top, with from ten to fourteen strong, sharp-edged, wavy ribs, the furrows also being wavy. Spine tufts set in little depressions along the margins of the ribs, five spines in each tuft, the two upper 1 in. long and four-angled, the two lower flattened and shorter, the fifth, which is the longest, being placed in the top of the cushion. Flowers 2 in. across, daisy-like, produced in April and May; tube very short; sepals and petals linear, spreading, white, with a purple stripe down the centre; stamens red, with yellow anthers; pistil purple, with an eight-rayed, yellow stigma. A native of Mexico. (Syn. *E. interruptus*.)

FIG. 30. — ECHINOCACTUS COPTONOGONUS

E. cornigerus (horn-bearing). — This remarkable plant, of which a portion is represented at Fig. 31, has the stoutest spines of all cultivated Cactuses, and their arrangement on the ridges of the stem is such as would withstand the attacks of all enemies. The broad tongue-like spine is purple in colour, and as strong as iron; the three erect horn-like spines yellow, and as firm as the horns of an antelope, to which they bear a resemblance. The stem is sphere-shaped, grey-green in colour, and is divided into from fourteen to twenty-one stout wavy ribs, upon which the spine tufts occur at intervals of about 2 in. The arrangement of the spines is shown in the illustration, as also is the position of the flowers, which are small, with narrow purple petals and brown-red sepals. The plant is a native of Mexico and Guatemala, and would require stove treatment. We have seen only small living examples, but according to descriptions and figures, the most interesting character it possesses is its spiny armament. It has been called *Melocactus latispinus* and *Echinocactus latispinus*.

FIG. 31.—ECHINOCACTUS CORNIGERUS

E. corynodes (club-like); Fig. 32.—The stem of this is about as large as a Keswick Codlin apple, with the broad end uppermost, and the sides cut up into about a dozen and a half rather prominent sharp ridges, with bunches of *stout* yellow spines arranged, at intervals of about 1 in., along the edges. The flowers, which are produced in a cluster on the top of the stem, form a crown of bright yellow petals, studded with scarlet eye-like stigmas. Each flower is 2 in. in diameter when fully spread out, cup-shaped, and composed of two or three rows of over-lapping petals. In the middle of these nestle the short stamens, and projecting well beyond them is the bright scarlet stigma, forming a beautiful contrast to the petals. This plant is a native of Mexico, and was introduced about the year 1837. It is also known in gardens under the names of *E. rosaceus* and *E. Sellowianus*. There was a pretty little specimen of this flowering in the Kew collection last year, and the opening and closing of the flowers, as the sunlight reached or receded from them, was almost as rapid as that observed in the daisy. The whole plant is so small, and, when in flower, so charming, that no one could fail to admire it. It requires similar treatment to *E. concinnus*.

FIG. 32.—ECHINOCACTUS CORYNODES

E. crispatus (curled); Fig. 33.—The flattened, wavy or curled ridges of this species are characteristic of several other kinds of Echinocactus. Its long, stout, ferocious-looking spines, directed upwards, have a very forbidding aspect. The stem grows to a height of about 8 in., and is said to produce its large, long-tubed, purple flowers in the summer months. It has been introduced by a Continental nurseryman, but, so far as is known, has not yet flowered in any English collection. It is apparently closely allied to *E. longihamatus*.

FIG. 33. – ECHINOCACTUS CRISPATUS

E. Cummingii (Cumming's); Bot. Mag. 6097. – A pretty little species, with a globose stem about 3 in. in diameter, the ridges divided into tubercles, and running spirally round the stem. From each tubercle springs a radiating cluster of yellowish, hair-like spines. The flowers are numerous, 1 in. long and wide, the scales on the tube tipped with red, whilst the petals stamens, and stigma are an uniform bright ochre-yellow; so that, looked at from above, they suggest the flowers of the common marigold. A well-managed plant produces as many as half-a-dozen of these flowers together, which open out widely under the influence of bright sunlight. It is one of the hardiest of the genus, thriving well in a frost-proof house or frame. During winter, the atmosphere surrounding it should be as dry as possible; but in summer it likes plenty of moisture, and exposure to full sunshine. A variety of *E. Cummingii* was raised from seeds a few years ago by Mr. Daniel, of Epsom, the flowers of which were pale almost to whiteness. The type is said to attain a height of 8 in. in its native country, Bolivia, whence plants were introduced to Kew in 1847, and flowered in July.

E. cylindraceus (cylindrical). – A large-growing kind, attaining a height of several feet, very broad, and, as the name denotes, cylindrical in shape. When large, the stem often develops lateral branches about its base. Cultivated plants of it are 6 in. high, the sides marked with about a score of ridges, upon which, arranged in a dense cluster, are the stout, strong spines, the longest of them 3 in. long, hooked, and projecting outwards, the shorter spreading and interlacing so as to form a sort of spiny network all round the stem. The flowers are yellow, 2 in. long, and are composed of a short, thick tube bearing from forty to fifty fringed sepals, and about half that number of petals, which are also fringed. There are as many as a dozen flowers opened together on stout, aged plants; it is, however, more because of the densely spinous stems than the flowers that this species has found its way into cultivation. It cannot be recommended for any except large collections, and where it can be grown in a stove temperature. It is a native of the hot deserts of Colorado, and was introduced about ten years ago. There are several healthy young specimens of it in the Kew collection.

E. echidne (viper; probably in allusion to the fang-like spines).—This species is remarkable in having a stout cylindrical stem, 12 in. high by 8 in. wide, with about a dozen deep ridges; these are disposed spirally, and bear tufts of rigid, broad spines, 1 in. or more long, spreading, so as to interlace and form a wire-like network all round the stem. It may be mentioned here that an American naturalist has recently suggested that the object of these iron-like spines on the stems of many Cactuses, and more especially on the majority of the Hedgehog kinds, is not so much to defend the fleshy stems from browsing animals as to afford protection from the scorching rays of the sun, which would otherwise cause the stems to blister and shrivel; and the nature of the spiny covering of *E. echidne* seems to support such a view. As in many others, the clusters of spines in this plant have their bases embedded in a tuft of whitish wool. The flowers are developed near the centre of the top of the stem, and are of medium size, bright yellow, with whitish stamens; they are produced two or three together, in summer. This species is a native of Mexico; it thrives in a greenhouse where frost is excluded, but only rarely flowers with us under cultivation.

E. Emoryi (Emory's); Fig. 34.—This is a very large-stemmed kind, specimens having been found nearly 3 ft. in height and about 2 ft. in diameter. Smaller plants, such as are in English collections, have globose stems 1 ft. through, with about thirteen ribs, the ribs tuberculated, the tubercles large, and rounded; the spines are borne on the apex of the tubercles in star-shaped bundles of eight or nine, and are angled, often flat on the top side, articulated, with hooked points, whilst in length they vary from 1 in. to 4 in. The flowers are 3 in. long, the tube clothed with heart-shaped scales or sepals; the petals are red, with yellowish margins, spreading so as to form a beautiful, large, cup-like flower, with a cluster of deep yellow stamens in the centre. The flowering period is in the autumn, and the plant is a native of the Lower Colorado and California. Living plants of it have only recently been introduced into English collections. At Kew, it is cultivated in a warm greenhouse, where it is in good health. From accounts of it in its native haunts, it will, however, probably prefer a cool house in winter, and as much sun and warmth as possible in spring and summer; for we are told that during winter it is often subjected to severe frosts and heavy snowfalls, whilst in summer the fierce heat of the sun is such as to burn up all vegetation, except Cactuses and other similar plants.

FIG. 34.—FLOWER AND SPINES OF ECHINOCACTUS EMORYI

E. gibbosus (humped).—A small apple-like plant, not more than 4 in. high, with a depressed top, the lower part being narrowed. It has sixteen ribs or ridges, composed of rows of thick fleshy tubercles, upon every other of which are six or eight horny spines, 1 in. long. The flowers are pushed out from the edge of the depression on the top of the stem, and are large; the tube 1= in. long. The petals spread to a width of 3 in., and are arranged in several rows, overlapping each other, becoming smaller towards the centre of the flower, as in an aster; they are pure white, except for a tinge of red on the tips of the outer ones, the stamens being bright yellow. Two flowers are usually developed on a plant, generally in June. This species was introduced from Jamaica about 1808, by a nurseryman in Hammersmith; but as no Echinocactuses are wild in the West Indian Islands, it must have been introduced into Jamaica from some of the Central American States, or probably from Mexico. It may be grafted on to another free-growing kind with advantage, as it does not always keep healthy when on its own roots. It should be grown in a cool greenhouse, or in the window of a dwelling-room, always, however, in a position where it would get plenty of sunlight.

E. Haynii (Hayne's); Fig. 35.—An upright cylindrical-stemmed species, very much like a Mamillaria in the form and position of the tubercles and the numerous greyish hair-like spines arranged in a radiating ring on the top of each tubercle. The flowers are much longer than in any yet described, the tube being 6 in. in length, clothed with large sepals on the upper portion, and the petals are semi-erect with recurved points, and coloured a brilliant purple-red. A native of Peru, where it is found at high elevations, growing in crevices of rocks and exposed to full sunlight. With us it thrives in a warm greenhouse, producing its beautiful flowers in summer. Introduced about 1850.

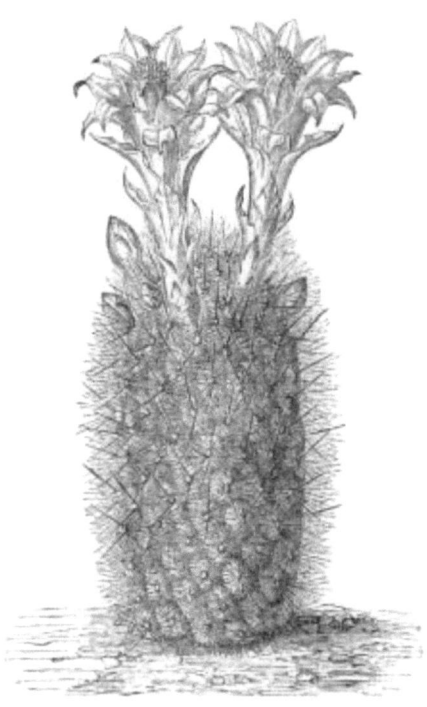

FIG. 35.—ECHINOCACTUS HAYNII

E. hexaedrophorus (tubercles six-sided); Fig. 36.—This plant is distinguished by the gouty-looking tubercles into which its broad, spiral ridges are divided, and which look as if they would suddenly burst like a blister if pricked with a pin. It grows about 4 in. in height, and is similar in form to what is shown in the accompanying figure, except that the top is usually flatter than here represented. The whole stem has a glaucous hue, and the spines are reddish-brown. The flowers, which are produced freely in June and July, are short-tubed, spreading to a width of 2 in.; the petals toothed at the tips, and arranged in several rows, overlapping each other, the colour being white, tinted with rose, with a disk-like cluster of yellow stamens in the centre, and a white-rayed stigma. A native of North Mexico, introduced about 1830. It is very slow-growing, attaining full size in not less than six years from seed; indeed, it is stated that in twelve years a plant of it did not grow more than 2 in. Still, slow as it is, it remains in good health when kept in a sunny position on a shelf in a greenhouse or in a dwelling-room, so that it may be recommended for places where space is very limited. Like *E. gibbosus*, it does best when grafted on to another kind. We have seen perfect "drum-sticks" formed by grafting a full-grown plant of this on the stem of a Cereus.

FIG. 36.—ECHINOCACTUS HEXAEDROPHORUS

E. horizonthalonis (spreading-spined); Fig. 37.—Stem globose, usually flattened on the top, and divided into eight or nine large ribs or ridges, grey-green in colour. Spines in crowded star-shaped clusters along the apex of the ribs, seven spines in each cluster, all of them strong, slightly curved, horn-like, and marked with numerous rings; they are yellow, tipped with red when young, ash-coloured when old; the longest are about 2 in. in length. Flowers terminal, springing from the young spine tufts, each 4 in. across, with two rows of petals arranged regularly in the form of a cup; colour deep rose, paler on the inside of the cup; stamens very numerous, with white filaments and yellow anthers. The flowers expand at sunrise and close again in the evening, each one lasting about a week; they are very agreeably scented. Flowering season, May and June. The plant is a native of Mexico, and was introduced in 1838. (Syns. *E. equitans, E. horizontalis*.)

FIG. 37.—ECHINOCACTUS HORIZONTHALONIS

E. Le Contei (Le Conte's); Fig. 38.—Another of the large-stemmed kinds, which have only recently found their way into English gardens, although long since discovered and described by American travellers. The illustration represents a young plant. When full-grown, this species has a stem 5 ft. high by 2 ft. wide, with broad deep channels and ridges, wrinkled and covered with a thick network of stout spines, which are set in clusters in a cushion of whitish wool, the longest being about 3 in. in length, with curved or slightly hooked points, and distinctly angular. The flowers are 2 in. long, bell-shaped; the petals shining lemon-yellow, with a tinge of brown on the outside, whilst the sepals are like a number of fish-scales, overlapping each other down the outside of the campanulate tube. The stamens and pistil are almost hidden inside the flower. Flowers are borne by quite young plants, whilst upon full-grown specimens they are so numerous as to form a large yellow cap to the immense, prickly, conical stems. They are developed in August and September. A native of Mexico, where it is found wild on the rocky or gravelly plains and ravines, and often in crevices of perpendicular rocks. It requires warm greenhouse treatment, and plenty of water during the summer, care being taken that the soil it is planted in is perfectly drained.

FIG. 38. — ECHINOCACTUS LE CONTEI

E. Leeanus (Lee's); Bot. Mag. 4184. — This species has many characters in common with *E. hexaedrophorus* and *E. gibbosus*, the stem being no larger than a small orange, with plump globose tubercles, bearing star-shaped clusters of short brown spines. The flowers are 1= in. long and wide, and are composed of a green fleshy tube, with a few whitish scales, which gradually enlarge till, with the white, rose-tipped petals, they form a spreading cup, the large cluster of pale yellow stamens occupying the whole of the centre. This pretty little Cactus was raised from seeds by Messrs. Lee, of the Hammersmith Nursery, in 1840. It is a native of the Argentine Provinces, and flowers in May. The treatment recommended for *E. gibbosus* will be found suitable for this. It is happiest when grafted on to another kind. For the amateur whose plants are grown in a room window or small plant-case, these tiny Hedgehog Cactuses are much more suitable than larger kinds, as they keep in health under ordinary treatment, and flower annually; whereas, the larger kinds, unless grown in properly-constructed houses, rarely blossom.

E. longihamatus (long-hooked); Fig. 39.—We heartily wish all species of Cactaceous plants were as readily distinguished and as easily defined in words as in the present remarkably fine and handsome one —remarkable in the very prominent ridges, the large and regularly-arranged spines, the central one very long, flattened, and usually hooked at the end, and handsome in the size and colouring of its flowers, both in the bud and when fully expanded. The stem is globose, 8 in. or more high; it has about thirteen prominent rounded ridges with waved tumid edges, from which, about 1= in. apart, spring clusters of spines, about a dozen in each cluster, dark red when young, becoming brown with age. In length, these spines vary from 1 in. to 6 in., the latter being the length of the central, hooked one, which is broad and flattened at the base. The flowers are 4 in. broad and long, the tube short, green, and bearing reddish scales, which gradually pass into bright yellow petals blotched with red on the outside, the inner ones spreading and forming a shallow cup, in the centre of which are the short yellow stamens and large pistil. Plants of this species have been grown with stems 20 in. high; but it takes a great number of years for the development of such specimens. The flowers are produced on the apex of the stem in July. This species was introduced from Mexico about 1850; it thrives only when grown in a warm greenhouse, where the temperature in summer may be allowed to run up to 90 degs. with sun heat. For large collections it is one of the most desirable.

FIG. 39.—PORTION OF PLANT OF ECHINOCACTUS LONGIHAMATUS

E. Mackieanus (Mackie's); Bot. Mag. 3561.—A small plant, not more than about 5 in. high, and 2 in. broad at the base, widening slightly upwards. The ridges are broken up into numerous fleshy, rounded, green tubercles, crowned with a tuft of thin brown spines from = in. to 1 in. long, their bases set in a small pad of yellow wool: As the stem gets older, it loses its tubercles at the base, which are changed into brown wrinkles. The flowers are developed on the top of the stem, generally two or three together, egg-shaped and scaly when in bud, 2= in. across when expanded; the petals white, tipped with brown; the stigma green, club-shaped. This curious little Cactus is one of about a dozen species found in the Chilian Andes. It was introduced in 1837 by the gentleman whose name it bears, and who, at that time, possessed a famous collection of Cacti. Like the rest of the Chilian kinds, it should be cultivated in a cool greenhouse in full sunshine, where it will produce its flowers in summer.

E. mamillarioides (Mamillaria-like); Bot. Mag. 3558.—This is another small, tubercled species, which, like the preceding, is a native of Chili. Its stem is very irregular in form, owing to the crowding of the tubercles, which look as if they were filled with water. The spines are small, in tufts of about half a dozen, set in a little cushion of yellowish wool. In size, the whole plant is like *E. Mackieanus*, but it blossoms more freely, as many as sixteen flowers having been borne at one time by a plant at Kew. These were short-tubed, the calyx clothed with green scales, and the petals 2 in. long, recurved at the apex, forming a beautiful cup-like flower of a bright yellow colour, with a band of red down the centre of each petal; the stamens and pistil yellow. The number of flowers developed on the small stem formed by this plant is quite extraordinary. It grows and flowers freely in an ordinary greenhouse, and would thrive in a sunny window if kept dry during the winter.

E. mamillosus (nipple-bearing).—A short, dumpy plant, with numerous tubercled ridges, bearing bunches of dark brown hair-like spines, which form a close network about the stem. The flowers are developed on the top of the stem, and are about 4 in. in diameter, with a thick tube; the petals are spreading, bright yellow in colour, and arranged in a regular, bell-like whorl. Inside this bell is a circle of purple filaments or stamens, forming a pretty contrast with the clear yellow of the petals. This is a recent introduction, which flowered in the Kew collection for the first time in June, 1886. It is one of the most beautiful of the large-flowered kinds, and, as it thrives in a warm greenhouse and is very free-flowering, it may be expected to become a favourite with Cactus growers. Owing to the lack of information respecting the conditions under which many of the Cactuses are found wild, and to the fact that little in the way of experimental culture has been done by growers of this family, cultivators are sometimes in the dark as regards the lowest temperature in which the rarer kinds can be safely grown. Many of the species of the present genus, for instance, were grown in stoves years ago but are now known to thrive in a cool greenhouse where frost alone is excluded.

E. multiflorus (many-flowered); Bot. Mag. 4181.—A well-named Cactus, as its small stem (seldom more than 5 in. high, and the same in width) often bears a large cap-like cluster of beautiful white flowers, except for a slight tinge of brown on the tips of the petals. Each flower is composed of a green, scaly tube, and several rows of reflexed petals, which form a shallow cup 2= in. across. The stamens are tipped with orange-coloured anthers, and the stigma is rayed and snow-white. The stem is ridged with rows of fleshy mammae or tubercles, which are curiously humped, and each bears a cluster of spreading, brown spines, 1 in. long. The number of flowers this little plant annually produces seems more than could be possible without proving fatal to its health; but we have seen it blossom year after year, and in no way has its health appeared impaired. It may be grown on a shelf in a warm greenhouse, or in the window of a heated dwelling-room. Introduced, probably from Mexico, in 1845. This, like all the small, globular-stemmed kinds, may be grafted on the stem of a Cereus of suitable thickness. Some cultivators believe that grafting causes the plants to flower more freely, but we have not observed any difference in this respect between grafted and ungrafted plants.

E. myriostigma. (many-dotted); Fig. 40.—In the form of the stem of this species we have a good illustration of how widely a plant may differ from others of the same genus in certain of its characters, for the spines are almost totally suppressed, and the ridges are regular, deep, and smooth. There are usually five or six ridges, a transverse section of the stem revealing a form exactly like the common star-fish (Astrophyton), a resemblance to which the name Astrophytum, sometimes applied to this plant, owed its origin. The form of the stem is well represented in the Figure. The white dots shown on the bark, and which look like scales, are composed of very fine interwoven hairs, which, under a microscope, are very pretty objects. This species was introduced from Mexico along with the large plant of *E. Visnaga* described at the beginning of this chapter, and was first flowered at Kew, in July, 1845. Stems 1 ft. in length were received, along with shorter ones; but only the small ones were established. The flowers are daisy-like, 1= in. across, and are straw-coloured, the petals being tipped with black. It thrives under warm greenhouse treatment. When without its flowers, it looks more like a piece of chiselled stone than a living plant.

FIG. 40.—ECHINOCACTUS MYRIOSTIGMA

E. obvallatus (fortified); Fig. 41.—The form of stem in this species is shown in the Figure. It grows very slowly plants 4 in. through taking about ten years to reach that size from seeds. The spines are stout, all deflexed, and arranged along the edges of the numerous ribs into which the stem is divided. The flowers are developed from the centre of the plant, and are surrounded by the erect spines, which crown the, as yet, undeveloped tubercles. Two or three flowers are produced at about the same time, each one being composed of a short, spiny tube, and a whorl of erect petals, which are pointed, purple-coloured, paler at the margin, the stamens being yellow. Native of Mexico. It requires a stove temperature, and flowers in summer.

FIG. 41.—ECHINOCACTUS OBVALLATUS

E. Ottonis (Otto's); Bot. Mag. 3107.—A dwarf kind, with a balloon-shaped stem, rarely exceeding 4 in. in height, and divided into a dozen wide ridges with sharp, regular edges, along which are clusters of small, brown spines, set in little tufts of wool, and looking like an array of spiders. The flowers are borne on the tops of the ridges, and are pale yellow in colour, with a band of red hair-like spines surrounding the calyx just below the petals, which are narrow, spreading, and look not unlike the flowers of the yellow Marguerite; the stigma is bright red. The symmetrical form of the stem, with its rows of spider-like spines, renders this plant attractive, even when without its bright and pretty flowers. It thrives only in a warm stove. Introduced from Brazil in 1831, flowering in the month of July. As it produces young plants about its base, it may be easily propagated by removing them and planting them in soil; or they may be grafted as advised for other of the small, globose-stemmed kinds.

E. pectiniferus (comb-bearing); Bot. Mag. 4190.—One of the most striking of the plants of this genus, owing to the character of its stem, and the large size and beauty of its flowers. The former resembles a pear with the thin end downwards; its height is from 4 in. to 6 in., and it has about twenty ridges, which are sharply defined and bear along their angles little cushions of white wool = in. apart, with a radiating cluster of brown spines springing from each. The arrangement of the spines in rows is not unlike the teeth of a comb. The flowers are borne near the top of the stem, and consist of a green, fleshy tube, clothed with spines and little tufts of white wool; the sepals form a row beneath the petals, and are yellowish, tinged with purple; petals 2 in. long, broad, with the upper margins toothed and the tip acute, their colour being bright rose, tinged with greenish-white at the base; stamens yellow; stigma large, green. The form of the flowers is that of a cup, nearly 3 in. across. Introduced from Mexico in 1845. Flowering season, April and May. It requires warm-house treatment.

E. polycephalus (many-headed); Fig. 42.—Stem globose when young, becoming cylindrical with age; number of ribs varying from twelve to twenty, sharply defined, and bearing, at intervals of 1 in., clusters of stout, reddish spines, somewhat flattened on the upper side, and marked with raised rings, or, as it is termed, annulated, the central ones attaining a length of over 3 in. on old plants, and sometimes curved. The flowers are enveloped at the base in a dense mass of white wool, which hides the tube, its spines only showing through; petals narrow, 1 in. long, spreading like a saucer, and coloured bright yellow; stamens numerous, yellow, as also is the large rayed stigma. California and Colorado, on stony and gravelly hills. Flowers in spring; introduced to Kew in 1886. This new plant is remarkable in that it is often found wild with as many as twenty to thirty stems or heads springing from the same base, and even young plants show early a disposition to develop several heads. The largest stems are from 1= ft. to 2= ft. high, and have a somewhat forbidding appearance, owing to the size and strength of their numerous spines. For its cultivation, a warm-house temperature appears most suitable; it bears a close resemblance to *E. texensis*.

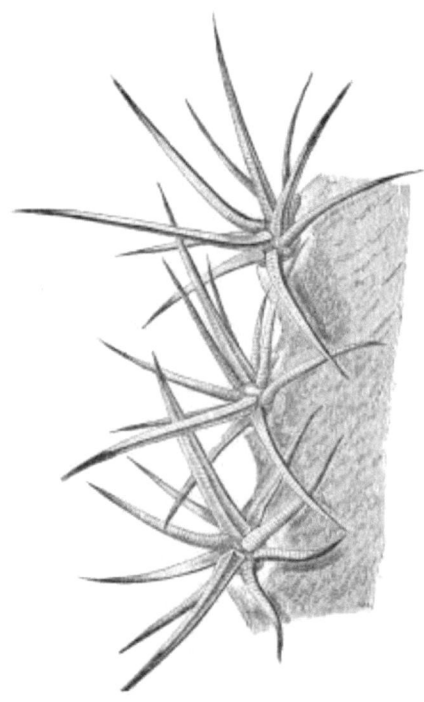

FIG. 42.—RIDGE, WITH SPINES, OF ECHINOCACTUS POLY-CEPHALUS

E. Pottsii (Potts').—The stem of this is shown in Fig. 43. Full-sized plants are 1= ft. in diameter, and have about a dozen ridges with acute sinuses, the ridges being rounded and even. The spines are 1 in. long, bristle-like, and are arranged in clusters of seven or nine, with a cushion of white wool at the base of each cluster. Flowers short-tubed, about 2 in. across, and coloured yellow; they are produced on the top of the stem in summer, several expanding together. The plant is a native of California, and was introduced about 1840. Under cultivation this species proves to be a shy-flowering Cactus, although in a warm house it grows freely, and remains in good health. It is well adapted for grafting on to the stem of some kind of Cereus, and in this way may be made to look very singular, as was shown in Mr. Peacock's collection of succulents some years ago, when a fine specimen, over 1 ft. across, was successfully grafted on to three stems of C. *tortuosus,* and had much the appearance of a melon elevated on a short tripod.

FIG. 43. — ECHINOCACTUS POTTSII

E. rhodophthalmus (red-eyed); Bot. Mag. 4486, 4634. — Stem cone-shaped, 4 in. to 1 ft. high, deeply furrowed; ridges about nine, 1 in. high, the angles bearing closely-set clusters of radiating spines, with a projecting one in the middle of each cluster, which contains nine spines 1 in. long, purple when young, becoming white when old. The flowers are produced from the summit of the stem, and have a thick, green, scaly calyx tube, upon which the spreading, rose-coloured petals are arranged in a regular series, and form a shallow bell nearly 3 in. across. The throat of the flower is coloured a deep crimson, against which the little sheaf of white stamens and the star-shaped yellow stigma form a pretty contrast. Three or more flowers are expanded together on a plant. It is a native of Mexico; introduced in 1845. It thrives in a house or frame where it is protected from frost, and during summer gets plenty of sunlight and air. It flowers in August. During the months of April and May, when it starts into growth, it should be kept close; but by the end of June, it should be exposed to the open air and allowed to ripen, so that its flowers may be produced in the autumn. The plant called *E. v. ellipticus* does not differ from the type, owing its name to the form of the stem of the first plant that flowered at Kew.

E. scopa. (brush-like); Fig. 44.—The stem of this species, when seen covered with numerous tufts of bristly spines, has been compared to a brush, a comparison not, however, applicable to the form represented in the Figure. In height the stems sometimes reach l= ft., with from thirty to forty ribs, bearing little discs of white wool at the bases of the clusters of spines. The flowers spring from the upper part of the stem (the nodules shown in the illustration represent the places where flowers have been developed at an earlier stage of growth), from four to six being borne in the same season; they are 1= in. long and wide, the tube short and brown, bristly; the petals are arranged in several overlapping series, rather wide for their length, toothed at the ends; their colour is a bright sulphur-yellow, as also are the stamens, whilst the stigma, which is rayed, is bright crimson. Native of Brazil. Introduced about 1840; it is more like a Cereus, in the form of its stem, than an Echinocactus. It flowers in June, and requires stove treatment. The stems, when dried carefully and stuffed with wadding, form pretty ornaments.

FIG. 44. — ECHINOCACTUS SCOPA

E. scopa cristata . (crested variety); Fig. 45. — This curious monstrosity owes its origin to fasciation similar to what occurs in the Celosias or Cockscombs, in some Echeverias, &c. These monster varieties of Cactuses do not flower, but they are nevertheless interesting, and worth growing on account of their curious shapes. The plant shown in Fig. 45 is grafted on the stem of a Cereus, and it is remarkable that a portion of the crest of the Echinocactus will, if grafted on to another plant, develop the abnormal form of its parent, proving that the variation, whatever its cause, has become fixed.

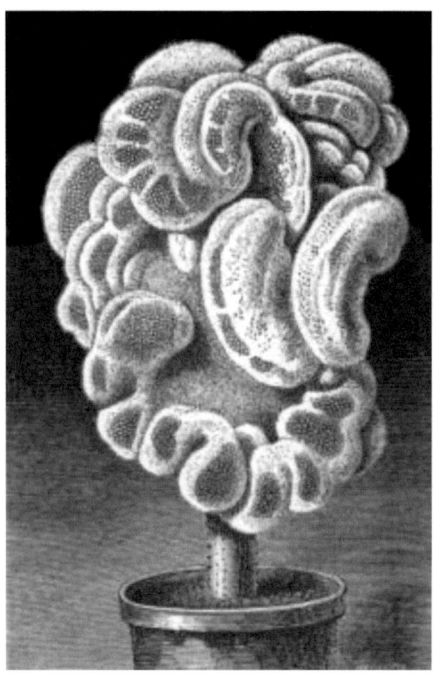

FIG. 45. — ECHINOCACTUS SCOPA CRISTATA

E. Simpsoni (Simpson's). — One of the smallest plants in the genus, and one of the prettiest. It produces tufts of irregularly-formed stems about 4 in. high, and composed of numerous rounded tubercles over = in. wide, bearing on the top of each a tuft of about twelve spines = in. long. The flowers are borne from the apex of the young tubercles, and are 1 in. wide and long, cup-shaped; petals pale purple, the stamens yellow. Native of Mexico and Colorado, where it is found at elevations of 8000 ft. to 10,000 ft., in great abundance, forming large patches on gravelly morains, where the climate during the summer is dry, whilst in winter a thick covering of snow protects the plants from severe frosts. In England, this species is said to have withstood 32 degs. of frost without being injured. It has been grown out of doors in a garden at Northampton, where it passed several winters planted in a raised border at the foot of a south wall with a natural coping of ivy. In New York, where the frosts of winter are severer than in England, it is cultivated out of doors. In this country it is apt to be injured by excessive moisture and fogs; but by protecting it with a handlight from November to March or April, this is overcome. If grown in pots, it should be kept in a position where it can enjoy all the sunlight possible.

E. sinuatus (undulated).—Stem about 8 in. wide and long; globose, bearing fourteen to sixteen ridges, the edges of which are wavy or undulated, the prominent points crowned with tufts of thin, flexuous, yellow spines, the longest 1= in., and hooked, the shorter > in., and straight. The stem of *E. longihamatus* is very similar to this. Flowers developed on the top of the stem; tube short, scaly, green; petals yellow, spreading, and forming a cup 3 in. across, which is greenish outside. A native of Mexico, where it flowers in April. A recently-introduced kind, not yet flowered in this country. It is described as being a distinct, large-flowered, handsome species.

E. tenuispinus (thin-spined); Bot. Mag. 3963.—Stem globular, depressed, with ridges and spines similar to those of *E. Ottonis*; indeed, by some these two are considered forms of the same species. In the number and size of the flowers, their colour and form, and the time of flowering, there is no difference between them. Native of Mexico (and Brazil ?).

E. texensis (Texan); Fig. 46.—A short-stemmed plant, with a thick, leathery skin and broad-based ridges of irregular form, crowned with tufts of stout horny spines, the central one much the longest, flattened at the base, and strong as steel. The flowers are produced near the centre of the top, from the tufts of whitish wool which accompany the spines on the young parts of the ridges. They are 2= in. long and wide; the tube short and woolly; the petals spreading, beautifully fringed, and rose-coloured. Native of Northeast Mexico, where it grows on stony hillsides in full exposure to sunshine, and where, during winter, it has to endure weather verging on to frost. It thrives in a greenhouse under cultivation. Like several of the stout-spined, robust-stemmed kinds, this may find favour as a garden plant because of the character of its stem, and the extraordinary strength of its large iron-like spines.

FIG. 46. — ECHINOCACTUS TEXENSIS

E. turbiniformis (top-shaped). — A very distinct dwarf kind, with globular stems 2 in. high and about 3 in. wide, clothed with spirally-arranged rows or ridges of tubercles, similar to those shown in the figure of *E. hexaedrophorus*, except that, in the former, there are no spines on the mature tubercles, although, when young, they have each a little cluster of fine spines. The flowers expand in June, several together, from the top of the stem; they are round, 1 in. across, the petals being numerous, pale yellow in colour, tinged with red on the outside. Introduced from Mexico, 1840. This curious little plant requires stove treatment, and thrives when grafted on the stem of some other kind. It is sometimes known as *Mamillaria turbinata*.

E. uncinatus (hooked); Fig. 47. — A small species, with oval stems when young, older plants becoming cylindrical, as shown in the accompanying Figure. The height of the largest plant does not exceed 6 in., so that, when wild, it is often hidden by the long grass in which it is frequently found on stony hillsides at high elevations, in Mexico. The ridges are broken up into large tubercles, upon each of which is a tuft of short straight spines, arranged in a circle, and a long hooked one springing from the centre, and often attaining a length of about 4 in. In old plants the spines are almost white, whilst in young ones they are purplish. The flowers are borne in a cluster on the apex of the stem, and are nearly 2 in. long, the tube being short and spiny, and the petals numerous, arranged in a cup, their colour dark purplish-red, the tips pointed; the stamens are yellow, with orange tips. The flowers expand only when the sun shines on them, closing up again in dull light, but opening again, and remaining fresh for about a week. Introduced in 1850. Flowers in March and April. This plant may be grown in a cool, sunny greenhouse, or window, requiring only protection from frost in winter, and in summer plenty of light, with a moderate amount of water. There are several varieties of it described, their differences being chiefly in the shape of the stem.

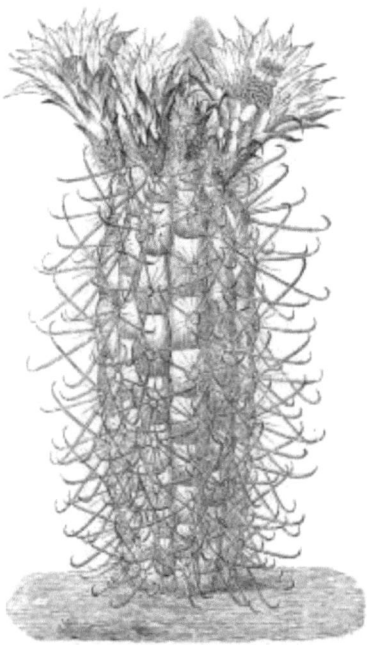

FIG. 47. — ECHINOCACTUS UNCINATUS

E. viridescens (greenish). — Stem 1 ft. high and 9 in. across, young plants being broader than high; the sides split up into about twenty ridges, which are again divided into knotty tubercles or waves. The spines are remarkable for their size and strength, those on large plants being 4 in. long by = in. broad at the base, gradually narrowing to a stiff point; there are four central spines of this size, the others, of which there are about a dozen, being shorter and thinner, and arranged stellately. The flowers, which are rarely produced, are poor in comparison with the majority of the flowers of this genus. As the name denotes, their colour is yellowish-green; and they are about 1= in. wide and high. There are often as many as a dozen flowers expanded together on a stem of this plant when wild, and they are arranged in a circle around the growing point. The interest in this species, however, centres in its spines rather than its flowers. It is a native of the dry hills of California, extending sometimes down to the sea-beach. There is a plant of it at Kew 6 in. high and about fifteen years old; it has not been known to flower there. Mr. Peacock also possesses a large plant of it.

E. Visnaga. (visnaga means a toothpick among the Mexican settlers); Fig. 48.—Of the most remarkable features of this truly wonderful Cactus we have already spoken earlier in this Chapter. In 1846, Sir W. J. Hooker described, in the *Illustrated London News*, a large plant of it, which had been successfully introduced alive to Kew, and which, a year or so later, flowered, and was figured in the *Botanical Magazine* (4559). Its height was 9 ft., and it measured 9= ft. in circumference; its weight a ton. Afterwards, it exhibited symptoms of internal injury. The inside became a putrid mass, and the crust, or shell, fell in by its own weight. The shape of the stem is elliptical, with numerous ridges and stout brown spines arranged in tufts along their edges. The flowers are freely produced from the woolly apex; the tube is scaly and brown, and the petals are arranged like a saucer about the cluster of orange-coloured stamens. The colour of the petals is bright yellow, and the width of the flower is nearly 3 in. This plant is a native of Mexico, and is usually cultivated in a tropical temperature, but it would probably thrive in a warm greenhouse. It flowers in summer. As we have stated, large specimens do not live long in this country; and as the flowers are only borne by such, the plant, except only for its stems, is not to be recommended for ordinary collections.

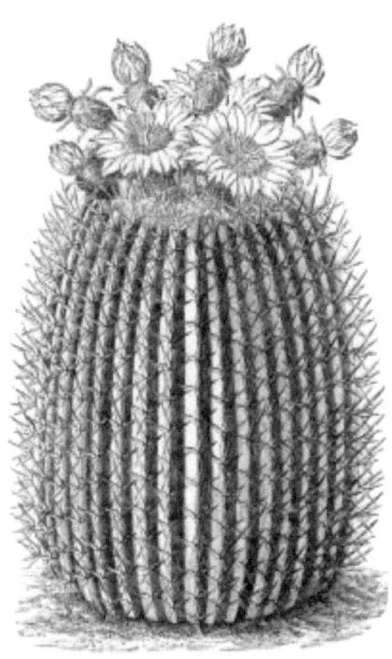

FIG. 48.—ECHINOCACTUS VISNAGA

E. Williamsii (Williams's); Bot. Mag. 4296.—A very distinct dwarf species, often called the "Dumpling Cactus," from the puffed-out, tumid appearance of its stems, which frequently branch at the base, so as to form a tuft of several heads; these are turbinate, 3 in. or 4 in. high, and 2 in. across the top, where the smooth, pale green flesh is divided into about half-a-dozen rounded tubercles, pressed closely together, and suggesting a number of small green potatoes joined by their bases. Each tubercle bears several tufts of short hairs. The flowers proceed from the young tubercles near the centre of the crown, their bases being enveloped in pale brown wool, the petals spreading out daisy-like to the width of 1 in., with a short disk of stamens in the middle; they are white, tinged with rose, and are developed in the summer months. Native of the rocky hills of Mexico, whence it was introduced in 1845. The stems of this plant are its most distinctive feature. It thrives on a shelf in a warm greenhouse, if kept perfectly dry in winter, and it should be potted in a compost consisting of broken brick two-thirds, loam one-third.

E. Wislizeni (Wislizen's); Fig. 49.—A large-stemmed kind, second only in size to *E. Visnaga.* Young plants have depressed stems, those in older specimens being cylinder-shaped. A specimen at Kew is 8 in. high by 18 in. in diameter, with twenty-one ridges, which are regular and sharp-edged, and bear bunches of spines at regular intervals, the outer and shorter ones being spreading and white, whilst from the middle of each tuft arise four longer and stouter spines, three of them 2 in. long, and one 3 in., with the point hooked, and as strong as if made of steel. The flowers, which are developed only on large plants, are greenish-yellow, about 2 in. long and wide, and expand during summer and autumn. The juice of the stems is said to serve as a substitute for water when the latter is scarce, and instances have been known among the white trappers where the lives of men have been saved by this plant. A novel use the stems are put to by the Indians is that of boilers, a purpose which they are said to answer well. The fleshy inside is scooped out, and the tough skin, with its iron-like spine protection, is then filled with vegetables and water and placed on the fire. As there is a plentiful supply of plants, the Indians do not trouble to carry this "boiler" about with them, but make a fresh one at every stage of their journeyings.

FIG. 49.—SPINES AND FLOWERS OF ECHINOCACTUS WISLIZENII

CHAPTER IX.

THE GENUS ECHINOPSIS.

(From *echinos*, a hedgehog, and *opsis*, like.)

N O less than three sections of Cactuses, viz., the above, Echinocactus, and Echinocereus, owe their names to their hedgehog-like stems. From a horticultural point of view, there is perhaps no good reason for keeping the above three genera and Cereus separate; but we follow Kew in the arrangement adopted here. The genus Echinopsis, as now recognised by most English botanists and cultivators, comprises about thirty species, most of which have been, or are still, in cultivation. They are distinguished from Echinocactuses by the length of their flower tube, from Cereuses by the form and size of their stems, and from both in the position on the stem occupied by the flowers. They are remarkable for the great size, length of tube, and beauty of their flowers, which, borne upon generally small and dumpy stems, appear very much larger and handsomer than would be expected.

The distribution of Echinopsis is similar to that of Echinocactus, species being found in Chili, Bolivia, Peru, Brazil, Mexico, &c. They grow only in situations where the soil is sandy or gravelly, or on the sides of hills in the crevices of rocks.

Cultivation.—The growing and resting seasons for Echinopsis are the same as for Echinocactus, and we may therefore refer to what is said under that genus for general hints with regard to the cultivation of Echinopsis in this country. The following is from the notes of the late Curator of the Royal Gardens, Kew (Mr. J. Smith), as being worthy the attention of Cactus growers. Writing about *Echinopsis cristata*, which he grew and flowered exceptionally well, he says: "This showy plant is a native of Chili, and, like its Mexican allies, thrives if potted in light loam, with a little leaf mould and a few nodules of lime rubbish. The latter are for the purpose of keeping the soil open; it is also necessary that the soil should be well drained. In winter, water must be given very sparingly, and the atmosphere of the house should be dry; the temperature need not exceed 50 degs. during the night, and in very cold weather it may be allowed to fall 10 degs. lower, provided a higher temperature (55 degs.) be maintained during the day. As the season advances, the plants should receive the full influence of the increasing warmth of the sun; and during hot weather, they will be benefited by frequent syringing overhead, which should be done in the evening. It is, however, necessary to guard against the soil becoming saturated, for the soft fibrous roots suffer if they continue in a wet state for any length of time."

None of the species require to be grafted to grow freely and remain healthy, as the stems are all robust enough and of sufficient size to take care of themselves. The only danger is in keeping the plants too moist in winter, for although a little water now and again keeps the stems fresh and green, it deprives them of that rest which is essential to the development of their large, beautiful flowers in summer.

SPECIES.

E. campylacantha. (curved-spined); Bot. Mag. 4567.—Stem 1 ft. or l= ft. high, globe-shaped, with a somewhat pointed top, the sides divided into from fourteen to sixteen ridges, with tubercled edges, bearing clusters of about ten strong brown spines, which are stellately arranged, a central one projecting outwards, then suddenly curving upwards, and measuring 3 in. in length. The flowers are developed from the ridges on the side of the stem; they are 6 in. long, the tube shaped like a trumpet, brownish in colour, and clothed with tufts of short black hairs; petals arranged in three rows, spreading so as to form a limb 2= in. across, pale rose-coloured, with a large cluster of yellow-tipped stamens, forming a disk-like centre. This species is a native of Chili, and was introduced in 1831. It blossoms in spring and summer. The long curving central spine and remarkable length of the flower-tube distinguish it from the other kinds. It may be grown in a cool greenhouse, where it will thrive, if kept freely watered during summer and rested on a dry, sunny shelf in winter. It is rare in English collections, but frequently occurs in Continental gardens.

E. cristata (crested); Bot. Mag. 4687.—Stem globe-shaped, 1 ft. high, slightly narrowed towards the top; ridges fifteen, 1 in. deep, sharply angular, the edges bearing tufts of spreading, yellowish spines, over 1 in. long, slightly curved, and tipped with red. Flowers creamy-white, springing from the ridges on the top of the stem; tube 4 in. long, clothed with tufts of black hairs, and surmounted by a whorl of reddish-yellow sepals, above which are two rows of broad-spreading petals. The width of the flower is over 6 in., and the stamens are arranged in a corona-like whorl inside the petals. This very fine Cactus is a native of Bolivia, whence it was introduced in 1850. When in flower, the broad, long-tubed, pale-coloured blossoms equal in beauty those of the Night-flowering Cereus. It blossoms in July. It thrives if kept in a warm, sunny greenhouse, but must be liberally treated in summer, so as to induce vigorous growth, and then be subjected to complete rest in winter in full sunlight, or it will not flower.

E. c. purpurea (purple).—This variety differs from the type in having deep rose-coloured flowers and a slightly longer tube. It is impossible to find among all the species of the Cereus section a more beautiful plant than this; the size of the flowers, their rich colour, their developing three or four together in the month of July, being almost exceptional, even among Cactuses. A splendid example of it was flowered at Kew in 1846 for the first time. It thrives under the conditions recommended for *E. cristata*. This variety is often made very sickly by treating it as a tropical Cactus, and, like most of these plants, if once it gets into a bad condition, it remains so a long time, in spite of liberal and careful treatment. So many of the Cactuses found in cool regions are ruined by an excess of heat in winter, and a close atmosphere during their season of growth, that too much attention cannot be given to the question of temperature in relation to their cultivation in English gardens.

E. Decaisneanus (Decaisne's).—As represented in Fig. 50, this plant appears to have a columnar stem, but this is owing to the specimen having been formed by cutting off the upper portion of an old plant and striking it. Naturally, the stem in this species is globular or slightly egg-shaped, and bears about fourteen ridges, upon which are tufts of short spines, springing from little cushions of whitish wool. The position of the flowers is shown in the figure. The tube is covered with tufts of hair-like spines, and the petals and sepals are broad, spreading, and white, tinged with yellow, as in *E. cristata*. The native country of this plant is not known; but it is a well-known garden Cactus, and thrives in a warm, airy greenhouse in summer, and on a dry, sunny shelf in winter. The swollen base of the tube is a good example of the nature of what is usually termed the flower-stalk in these plants. It is, as has been pointed out, the elongated calyx, and the swollen portion is the ovary or seed vessel. If, therefore, seeds are desired, the withering flowers should be allowed to remain, and, in time, the upper portion of the tube will fall away, leaving the base, which continues to grow till it attains the proportions of a hen's egg.

FIG. 50.—ECHINOPSIS DECAISNEANUS

E. Eyriesii (Eyries').—Stem no larger than an orange, with about a dozen ridges, the edges sharp, and bearing little globular tufts of whitish wool and red, hair-like spines. Flower exceedingly large for the size of the stem, the tube being more than 6 in. long, funnel-shaped, pale green, with tufts of brown hairs, which look very much like insects, scattered over the surface. The petals are numerous, narrow-pointed, spreading, pure white, the stamens pale yellow, and the star-like stigma white. This species is a native of Mexico, and was introduced by the late Sir John Lubbock, about 1830. It blossoms at various seasons, generally in summer. "Independently of the large size of the flowers, which rival in dimensions those of the Cereuses, it is remarkable for the rich, delicate odour they exhale at night, at which time its glorious blossoms expand. When young, they resemble long, sooty-grey horns, covered over with a thick, shaggy hairiness, and would never be suspected to conceal a form of the utmost beauty and a clear and delicate complexion.

When the hour of perfection has arrived, and the coarse veil of hair begins to be withdrawn by the expansion of the unfolding petals, one is amazed at the unexpected loveliness which stands revealed in the form of this vegetable star, whose rays are of the softest white" (Lindley). For its cultivation, this plant requires a warm house always; but care should be taken to give it plenty of fresh air and as much light as possible. The soil best suited for it is a rich loam with a little sand and charcoal. It likes liberal watering in summer.

E. E. flore-pleno (double-flowered); Fig. 51.—A form with several rows of petals, which give the flowers a doubled appearance.

FIG. 51.—ECHINOPSIS EYRIESII FLORE-PLENO

E. E. glauca (hoary-grey). This variety differs from the type in the absence of the dark brown hairs from the flower-tube, which is also shorter than in *E. Eyriesii*. Probably a native of Mexico.

E. oxygonus (sharp-angled).—This is very similar to *E. Eyriesii*. Stem globular in shape, and divided into about fourteen acute-edged ridges, upon which are tufts of brown spines, varying from = in. to 1= in. in length. Flower 8 in. long, the tube slightly curved, covered with little scales and hairs, and coloured green and red. The petals form an incurved cup, and are broad, with pointed tips; their colour a bright rose, with a lighter shade towards the centre of the flower. As in *E. Eyriesii*, the flowers of this kind are borne several together from the ridges near the growing centre of the stem. It is a native of Brazil, whence it was introduced nearly half a century ago. It thrives in an intermediate house, if treated as advised for *E. Eyriesii*, and its flowers will develop in summer. The extraordinary size and beauty of the blossoms are sufficient to compensate for their comparatively short duration after expanding; it is also interesting to watch the gradual development of the tiny, hairy cone, which is the first sign of the flower, and which increases in length and size at a surprising rate.

E. Pentlandi (Pentland's); Fig. 52.—A pretty little species, with a globose stem 3 in. in diameter, divided into about a dozen rounded ridges, which are undulated or broken up into irregular tubercles, when the ridges do not run parallel with each other. Each tubercle is crowned with a tuft of brown, bristle-like spines, = in. or so long. The flowers are large in proportion to the size of the plant, the tube being 4 in. long, and trumpet-shaped; petals arranged in several overlapping rows and forming a cup 2 in. across, the lowest whorl turning downwards; in colour, they are a brilliant red, the stamens white, and the stigmas yellow. Three or four flowers are often expanded together on the same stem, springing from the side instead of the top of the plant. Native of Mexico (?); introduced about 1840. There are several distinct seedling or hybrid forms of this species, remarkable in having the colour of their flowers either red, yellow and white, or white, whilst some, such as the one known as *flammea*, have flowers only 2 in. long. These kinds may all be grown in a sunny greenhouse or window, as they only require protection from frost. They may be placed out of doors in summer, and be kept under glass only during winter, treatment which will result in better growth and more flowers than if the plants were kept permanently under glass.

FIG. 52.—ECHINOPSIS PENTLANDI

E. P. longispinus (long-spined); Fig. 53.—This is a long-spined form, and differs also in the shape of the stem, which is oblong, rather than globose.

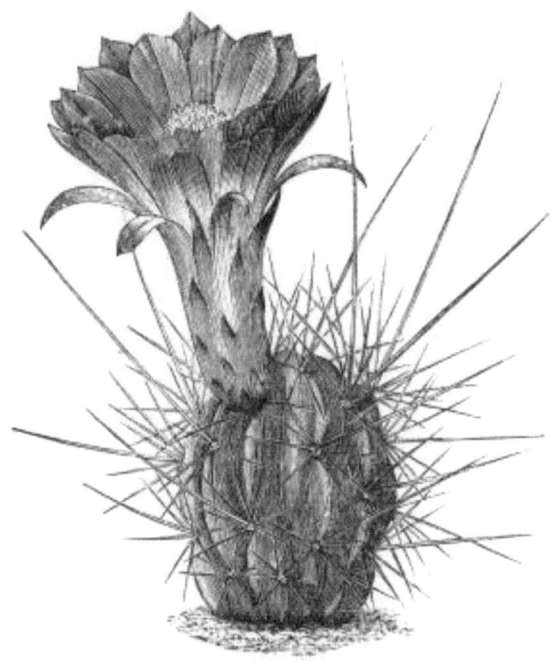

FIG. 53.—ECHINOPSIS PENTLANDI LONGISPINUS

E. tubiflorus (tube-flowered).—This species has an orange-shaped stem, about 4 in. high, and divided into about twelve prominent, sharp-angled ridges, along which are tufts of blackish spines, = in. long, and set in little cushions of white wool. The flower springs from the side of the stems, where it replaces a tuft of spines, and, as in *E. Eyriesii*, the tube is remarkably long, whilst the size of the whole flower much exceeds that of the rest of the plant, the length of the tube being about 6 in., and the width of the flower over 4 in. The petals are pure white, recurved, displaying the crown of yellow stamens, arranged in a ring about the rather small, rayed stigma. The tube is uniformly green, except that the scale-like bracts are edged with long, blackish, silky hairs. A native of Mexico; introduced about fifty years ago, when it was figured in the *Botanical Magazine* and elsewhere as a species of Echinocactus. *E. tubiflorus* may be placed along with *E. Eyriesii* and *E. oxygonus*, as it requires similar treatment. The three kinds here mentioned may be recommended as a trio of very fine-flowered, small-stemmed Cacti, which may be grown successfully in any ordinary greenhouse.

CHAPTER X.

THE GENUS MELOCACTUS.

(From *melon*, a melon, and *Kaktos*, a name applied by Theophrastus to a spiny plant; the species are melon-formed, and their angles are beset with tufts of spines.)

HIS genus forms a group of well-marked and curious plants, with stems similar to those of the globose Echinocactuses and floral characters quite distinct from all other genera. They cannot be said to possess any particular beauty, as their stems are stiff and dumpy, their spines large and rigid, and their flowers small and unattractive. But what is wanting in beauty of form or colour is atoned for in the cap which crowns the stem, and forms the flower-head, growing taller and taller whilst the stem remains stationary, till, under favourable circumstances, a cylindrical mass of spines and hairs, not unlike a large bottle-brush, and 1 ft. or more in length, is developed before the whole plant succumbs to old age. This character belongs more particularly to *M. communis*, the commonest species, and the one best known in English gardens. Additional interest attaches to this species, from the fact of its having been the first Cactus introduced into Europe, for we are informed that in the year 1581 living plants of the Melon Cactus were known in London. Fifty years later, Gerard, the Adam of English gardening, wrote: "Who can but marvel at the care and singular workmanship shown in this Thistle, the *Melocarduus echinatus*, or Hedgehog Thistle? It groweth upon the cliffes and gravelly grounds neere unto the seaside in the islands of the West Indies, called St. Margaret's and St. John's Isle, neere unto Puerto Rico, and other places in these countries, by the relation of divers that have journied into these parts who have brought me the plant itself with his seed, the which would not grow ill my garden, by reason of the coldnesse of the clymate." After this, the plant appears to have been frequently cultivated in gardens in this country, and it has only been in recent years that this and similar curiosities have almost disappeared from all except botanical collections.

The most prominent distinctive characters of Melocactus reside in the cap or cluster of spines, wool, and flowers on the summit of the stem. Thirty species are included in the genus, their stems ranging from 1 ft. to 3 ft. in height, the ridges straight, and, as a rule, large; whilst all have stiff stout spines in clusters about 1 in. apart. The small flowers are succeeded by bright red, cherry- like berries, containing numerous black, shining seeds. The distribution of the species is over the hottest parts of some of the West Indian Islands and a few places in Central and South America.

Cultivation.—The cultivation of the several kinds known in gardens is as follows: A tropical temperature all the year round, with as much sunlight as possible, and a moist atmosphere for about three months during summer, when growth is most active. Very little soil is required, as the largest stems have comparatively few roots; indeed, imported stems have been known to live, and even make growth, nearly two years without pushing a single root; but, of course, this was abnormal, and was no other than the using-up of the nourishment stored up in the stem before it was removed from its native home. M. Louis de Smet, a well-known Ghent nurseryman, who grows a fine collection of Cactuses, stated that he had kept *M. communis* a long time in robust health and growth by feeding it with a very weak solution of salt. Tried at Kew, this treatment did not appear to make any perceptible difference; but, bearing in mind that the Turk's-Cap Cactus is found in great abundance within the reach of sea spray, in some of the West Indian Islands, there seems much reason in M. de Smet's treatment. The same gentleman informed us that he had a specimen of this Cactus bearing no less than thirteen heads. There is, at the time of writing, a specimen at Kew bearing four fine heads. Large imported plants are very rarely, established; and even when established, they do not thrive long, owing to the fact that, after the cap has commenced to form, no further stem-growth is made. Young plants grow very slowly, a plant 3 ft. across taking, according to Sir W. Hooker, from 200 to 300 years to reach that size. It has been stated that grafting is a good plan to adopt for the Melocactus, Mr. F. T. Palmer, in "Culture des Cacties", recommending the following treatment for *M. communis*: Take a *Cereus peruvianus* of about the same diameter as that of the base of the Melocactus, cut off the head of the former, but not so low as to come upon the hard, ligneous axis, and then pare off the hard epidermis and ribs for about 1 in. Then take off a slice from the base of the Melocactus, also paring off about 1 in. of the epidermis all round; place the two together, and bind on firmly with strong worsted. In warm weather, a union should take place in about two months, but it will be safest to allow the ligature to remain till growth commences. The precaution of paring off the hard skin and ribs is absolutely necessary, as the juicy centre contracts, and the rind, or epidermis, does not. There would, therefore, be a cavity formed sufficient to prevent all cohesion, be the graft tied on ever so tightly.

Large imported stems should be kept perfectly dry for about a fortnight, and, if they show any signs of rottenness, they should be carefully examined and the bad portions cut away; exposure to the air for a few days will generally cause these pared places to callus over. At all times, even when the stems appear to be in good health, a sharp look-out should be kept for patches of rottenness in the stem, and especially about its base.

Propagation.—This is effected by means of seeds, which usually follow quickly after the flowers produced on cultivated specimens. Multiplication is also possible by means of offsets, which are formed about the base of the stem if the top of a growing plant is cut out. The thirteen-headed plant mentioned above was the result of the removal of the top of a stem which had developed these lateral growths, and thus formed a family of red-capped stems; this had, however, taken place before the plant was removed from its native home. As the cap is the most remarkable part of *M. communis*, the purchase of large imported stems, in preference to young ones raised from seeds, is recommended; for, as the cap does not form till the stem attains a large size, there would be small hope of seedlings reaching the flowering stage during a lifetime.

SPECIES.

M. communis (common); Fig. 54.—Stem from 2 ft. to 3 ft. in diameter, globose, with from twelve to twenty ridges, and armed with numerous clusters of strong, short spines, the clusters placed closely together. On the summit of the stem is a cylindrical crown, about 4 in. broad, and varying in height from 5 in. to 12 in. This cylinder is composed of a thick pad of whitish, cotton-like substance, through and beyond which a great number of bristle-like red spines are developed, the whole being not unlike a bottle-brush. About the top of this brush-like growth the flowers are produced. These are small, red, fleshy, and tube-shaped, the calyx and corolla forming a regular flower, as in a Hyacinth. They are borne at various times in the year, as long as the cap is growing; afterwards the latter falls off; and the stem rots. We have a cap that was cast by an old plant, and which has stood as an ornament on a shelf in a room for about four years, and is still in perfect condition. In addition to the name of Turk's-Cap Cactus this plant is also known as "Englishman's Head" and "Pope's Head." It is a native of several of the islands of the West Indies, being very abundant in St. Kitt's Island, where it grows in very dry, barren places, often on bare porous rocks.

FIG. 54.—MELOCACTUS COMMUNIS

M. depressus (flattened); Bot. Mag. 3691.—Stem broader than high, deeply cut into about ten broad furrows, along the sharp angles of which are clusters of pale brown spines, from = in. to 1 in. long, arranged in a star, each cluster 1 in. apart. Instead of the cylinder-like cap of the Turk's-Cap species, this one has a short, broad tuft of white wool and red spines, like a skull-cap. The flowers are small, and soon wither, but remain attached to the oblong berries, which stand erect in a dense cluster in the centre of the cap, and are of a delicate rose-colour. The first introduced plant of this was sent home by Mr. Gardner, who introduced the Epiphyllums and other Cactuses. It flowered on the way to England, and matured its seeds soon after its arrival. It is a native of Pernambuco.

M. Miquelii (Miquel's); Fig. 55.—This species appears to have been introduced in 1838, when two plants of it were sent from the West Indian Island, St. Croix, to the Hamburg Botanic Gardens. The stem is oval, dark green, with fourteen well-defined ribs, as regular as if they had been carved with a knife. The spine-tufts are small;

spines short, black-brown, about nine in each tuft, one of which is central, the others radiating; they are less than = in. long. The "cap" is cylindrical, 3 in. high by 4 in. in diameter, and composed of layers of snow-white threads, mixed with short reddish bristles.

FIG. 55.—MELOCACTUS MIQUELII

These three are the only species of Melocactus that have become known in English gardens, although various other kinds, named *M. Lehmanni, M. Zuccarini, M. Ellemeetii, M. Schlumbergerianus*, &c., occur in books.

CHAPTER XI.

THE GENUS PILOCEREUS.

(From *pilos*, wool, and *Cereus*, in allusion to the long hairs on the spine cushions, and the affinity of the genus.)

NE of the most striking plants in this order is the "Old Man Cactus," botanically known as *Pilocereus senilis*, which is the only member of this genus that has become at all known in English gardens. In Continental gardens, however, more than a dozen species are to be found in collections of succulent plants; and of these one of the most remarkable is that represented at Fig. 56. The limits of the genus Pilocereus are not definitely fixed, different botanists holding different views with respect to the generic characters. Recent writers, and among them the late Mr. Bentham, sunk the genus under Cereus; but there are sufficiently good characters to justify us in retaining, for garden purposes, the name Pilocereus for the several distinct plants mentioned here. The botanist who founded the genus gives the following general description of its members: Stems tall, erect, thick, simple or branched, fleshy, ridged; the ridges regular, slightly tubercled, and placed closely together. Tubercles generally hairy, with bunches of short spines; the hairs long and white, especially about the apex of the stem, where they form a dense mass. Flowers on the extreme top of the matured stems, and arranged in a cluster as in the Melon Cactus, small, tubular; the petals united at the base, and the stamens attached to the whole face of the tube thus formed, expanding only at night, and fading in a few hours. These flowers have a disagreeable odour, not unlike that of boiled cabbage. Fruit fleshy, round, persistent, usually red when ripe. The species are natives of tropical America, and are generally found in rocky gorges or the steep declivities of mountainous regions.

Cultivation.—These plants require distinctly tropical treatment. During summer, they must have all the sunlight possible, and be supplied with plenty of water, both at the root and by means of the syringe. Air should be given on very hot days, but the plants should be encouraged to make all the growth possible before the approach of winter. In winter, they may be kept quite dry, and the temperature of the house where they stand should be maintained at about 60 degs., rising to 65 degs. or 70 degs. in the day. In March, the plants should be repotted into as small pots as convenient, employing a good, loamy soil and ample drainage. Should the hairs become soiled or dusty, the stems may be laid on their sides and then syringed with a mixture of soft soap and warm water, to be followed by a few syringefuls of pure water; this should cleanse the hairs and give them the white appearance to which the plants owe their attractiveness.

SPECIES.

P. Houlletianus (Houllet's); Fig. 56.—Stem robust, glaucous-green; ridges about eight, broad, prominent, obscurely tubercled; spines in bundles of nine, radiating, straight, less than 1 in. long, and pale yellow. Upon the growing part of the stem, the spines are intermingled with long, white, cottony hairs, often matted together like an unkempt head; these hairs fall off as the stem matures. Flowers funnel-shaped, resembling Canterbury Bells, borne in a cluster on the summit of the plant; ovary short and scaly; petals joined at the base, and coloured a rosy-purple, dashed with yellow; the stamens fill the whole of the flower-tube and are white; style a little longer than the flower-tube, and bearing a ray of about a dozen stigmas. Fruit globose, as large as a plum, and coloured cherry-red. The pulp is bright, crimson, and contains a few brownish seeds. In the engraving the fruit is shown on the left, and a flower-bud on the right. This species is often known in Continental collections as *P. Fosterii*.

FIG. 56.—PILOCEREUS HOULLETIANUS

P. senilis (Old-Man).—Stem attaining a height of 25 ft., with a diameter of about 1 ft.; ridges from twenty-five to thirty on plants 4 ft. high; the furrows mere slits, whilst the tufts of thin, straight spines, 1 in. long, which crown each of the many tubercles into which the ridges are divided, give young stems a brushy appearance. About the upper portion of the stem, and especially upon the extreme top, are numerous white, wiry hairs, 6 in. or more long, and gathered sometimes into locks. To this character, the plant owes it name Old-Man Cactus; but, by a curious inversion of what obtains in the human kind, old plants are less conspicuous by their white hairs than the younger ones. Some years ago, there were three fine stems of this Cactus among the cultivated plants at Kew, the highest of which measured 18= ft. There was also, however, a fine specimen in the Oxford Botanic Gardens, with a stem 16 ft. high; and it is stated that this plant has been in cultivation in England a hundred years at least. A plant twenty-five years old is very small, and, from its slowness of growth, as well as from the reports of the inhabitants of Mexico, where this species is found wild, there is reason to believe that a stem 20 ft. high would be several hundred years old. The flowers of *P. senilis* are not known in English collections, the plant being grown only for its shaggy hairiness.

Other species are: *P. chrysomallus*, which has a branching habit, *P. Br|nnonii* (Fig. 57), *P. Celsianus, P. columna, P. tilophorus,* known only in a young state, and several others, all very remarkable plants, but not known in English collections, unless, perhaps at Kew.

FIG. 57. – PILOCEREUS BR\NNONII

CHAPTER XII.

THE GENUS MAMILLARIA.

(From *mamilla*, a little teat; in allusion to the tubercles.)

SOMETHING over 300 different kinds of Mamillaria are known, but only a small proportion of these may be considered as garden plants. They are characterised generally by short, symmetrically-formed stems, sometimes aggregated together and forming a dense tuft, but, as a rule, each plant has only one stem. The generic name is descriptive of the chief feature in these stems, namely, the closely-set, spirally-arranged tubercles or mamillae, which vary considerably in the different kinds, but are always present in some form or other. Some kinds have stems only 1 in. high by 2/3 in. in diameter, and the tubercles hidden from view by the star-shaped cushions of reddish or white spines. In some, the spines are erect and hair-like, giving the plant the appearance of tiny sea-urchins; another group has the principal spines hooked at the tip, and the points in these so sharp that if the hand comes in contact with them they hook into it and stick like fish-hooks. The purpose of these hooked spines seems doubtful; certainly, they cannot serve as any protection to the plant itself, as they are so strong that the plant must be torn up by the roots before the hooks will give way.

The spines in *M. macromeris* are straight, and measure 2 in. in length; in *M. multiceps* they are in two series, the one fine, white, and short, the other yellow and stout. The most marked section of this genus, however, is that represented by *M. fissurata* (Fig. 61), in which the tubercles are large, spreading horizontally, and angular, resembling most closely the foliage and habit of some of the Haworthias. No one who had not read up the botany of Mamillarias would suspect that this plant belonged to them, or even to the Cactus order at all. There is a good specimen of it in the Kew collection. When in flower the family resemblance is easily seen; but as this species does not flower freely, it will be known by its remarkable foliage-like tubercles, rather than as a flowering Cactus. And the same remark applies to many of the Mamillarias; their stems thickly beset with tubercles and spines, always regular in arrangement, and neat and attractive in appearance.

The following remarks made by Dr. Lindley when describing *M. tenuis*, give a good idea of the singular, yet pretty, stems of some of these plants: "Gentle reader, hast thou never seen in a display of fireworks a crowd of wheels all in motion at once, crossing and intersecting each other in every direction; and canst thou fancy those wheels arrested in their motion by some magic power—their rays retained, but their fires extinguished and their brightness gone? Then mayst thou conceive the curious beauty of this little herb—a plant so unlike all others that we would fain believe it the reanimated spirit of a race that flourished in former ages, with those hideous monsters whose bones alone remain to tell the history of their existence." It is quite true that in the cultivated Mamillarias there is nothing unsightly, or rough, or unfinished. Without foliage, their stems globose, or short cylinders, or arranged in little cushion-like tufts, and enveloped in silky spines, like tiny red stars, always looking the same, except when in flower, and never looking in the least like ordinary plants. Characters such as these ought to find many admirers. In the Succulent House at Kew, there is a long shelf upon which a great many plants of this genus may be seen. But the flowers in some of the species of Mamillaria are quite as attractive as the stems. Those of *M. macromeris* are 3 in. long and wide, their colour a deep rose; *M. Scheerii* has equally large flowers, and coloured bright yellow, as also are the flowers of *M. pectinata*. This last is remarkable on account of the clock-like regularity with which its flowers expand. While fresh, they open every day between eleven and twelve o'clock, and close again about one, however strong the sunlight shining upon them may be. Some of the kinds (more especially the small-flowered ones) are often prettily studded over with bright red, coral-like berries, which are the little fruits, and contain, as a rule, matured seeds capable of reproducing the parent plant.

The headquarters of the genus Mamillaria is Mexico, and the countries immediately to the north, a few being scattered over the West Indies, Bolivia, Brazil, and Chili. Many of them grow on mountains where the temperature is moderate, but where the sunlight is always intense. Others are found on limestone or gravelly hills, among short herbage, or on grassy prairies. A small silvery-spined kind has recently been found near the snow line in Chili. *M. vivipara* is quite hardy in New York, as also are several other kinds, whilst we learn that by planting them out in summer, and protecting them by means of a frame from heavy rain, dews, fogs, and sudden changes of weather, a good many species of both Mamillaria and Echinocactus are successfully managed in the neighbourhood of that town.

Cultivation.—Particulars with respect to cultivation are given along with the descriptions of most of the species, but a few general principles may here be noted. With only a few exceptions, all the cultivated Mamillarias may be grown in a warm, sunny greenhouse, or they may be placed in a frame with a south aspect, during our summer, removing them into artificially heated quarters for the winter. They do not like a large body of soil about their roots, but always thrive best when in comparatively small pots. If a sweet, new, fibry loam, mixed with broken bricks or cinders, be used to pot these plants in, they may then be left undisturbed at the root for several years. Much harm is often done to the more delicate kinds of Cactuses by repotting them annually; the best-managed collection I have seen had not been repotted for four years. This would not be safe if a poor and exhausted soil were used in the first instance. The pots should be well drained with crocks, and these covered with a layer of fibre sifted from loam. In summer, the soil should be kept moist, but never saturated; and after a bright warm day, the stems may be moistened over by syringing them with tepid water. A point of much importance in connection with these, and indeed all tropical and extra-tropical plants, is, that the water used for watering or syringing them should be rain-water if possible, and never more than a degree or so colder than the plants themselves would be. Thus, a plant which had been standing in the full glare of a midsummer sun all day, would be much endangered by watering it with cold tap-water. Where proper arrangements for water are not made in a greenhouse or stove, it is a good plan to place the water wanted for the day's use in the sun along with the plants. A little bag filled with soot and tightly tied at the neck, and water, is a good method for rendering hard tap-water suitable for watering the roots of plants. In winter, Mamillarias may be kept quite dry at the roots, except in mild sunny weather, when a little water may be given.

A collection of the most distinct kinds may be successfully managed in a glass case in a room window, providing the sun shines through it for a few hours in the day.

Propagation.—This is usually effected by means of seeds, which may be procured from Continental seedsmen as well as from our own. The treatment required by the seeds is similar to what has been already advised for those of other Cactuses. The tufted kinds are easily multiplied by separating the stems, or even by cutting off the tops and planting them in small pots of sandy soil.

SPECIES.

The following kinds are selected from those known to be in cultivation; of course, it is out of the question here to enumerate all the species known.

M. angularis (angular-tubercled).—A robust kind, with stems 4 in. to 8 in. high, and branching somewhat freely; tubercles prism-shaped, rather thick at the base, and slightly angular, < in. long, their tops tufted with short white spines; at the base of the tubercles are little tufts of white wool. Flowers are only rarely produced by cultivated plants; they are small, tubular, rosy-purple, the stamens yellow. Introduced from Mexico in 1835; flowers in summer. When happily situated, it forms a specimen 1 ft. in diameter, owing to its freely produced arm-like branches, which spread out and curve upwards. It requires a warm greenhouse temperature during winter, and exposure to bright sunshine at all times.

M. applanata (flattened). —In this, the stem is broader than high, and has a squat appearance; tubercles > in. long, cone-shaped, with stellate tufts of straight, hair-like spines, white when young, yellowish when aged. Flowers springing from the outside of the stem-top, white, tinged with red. It is a native of Mexico, and blossoms in summer. A specimen, 6 in. through at the base, may be seen at Kew, where it has been for many years, without altering perceptibly in size. All the larger-stemmed Mamillarias are exceedingly slow growers after they have reached a certain size, although, in the seedling stage, they grow freely. The treatment for this kind should resemble that advised for the last.

M. atrata (blackened).—Stem oval in shape, broad at the base, 4 in. high, unbranched; tubercles swollen, = in. long, deep green, cone-shaped, becoming flattened through pressure of growth. Spines set in a tuft of white hairs, falling off from the lowest mammae, as happens in many of the thick-stemmed kinds. Flowers numerous, and developed all round the outside of the stem, stalkless, nestling closely between the tubercles, and when expanded looking like starry buttons of a rosy-pink colour. Native of Chili, flowering in autumn. This species is rare in England, but is worth attention because of the prettiness of its flowers, the attractive form of its stem, and its reputed hardiness. It will thrive in a cold frame, and requires protection from excessive wet only, rather than from cold. Grown in a warm house, it becomes sickly, and is short-lived.

M. bicolor (two-coloured). — One of the commonest of the Cactuses grown in English gardens, and one of the most distinct, owing to its short, silvery hair-like spines, thickly crowded on the ends of the small tubercles, completely hiding the stem from view. The latter is from 6 in. to 1 ft. high, 3 in. in diameter, cylindrical, often branching into several thick arms, when it has a quaint appearance. If kept free from dust, which may be done by covering the plant with a bell glass, there is much beauty in the stem; indeed, it is owing to this, rather than as a flowering Cactus, that this species finds favour as a garden plant. The flowers are less than 1 in. in length and width, stellate, their colour deep purple; they are developed in June. Although a native of elevated regions in Mexico (4000-5000 ft.), this plant thrives best when grown in a warm house. There are several handsome and very old specimens of it in the tropical collection of succulents at Kew. It is one of the easiest to manage, and will thrive in a warm room-window if exposed to bright sunlight and kept dry in winter. *M. nivea* and *M. nobilis* are both varieties of this species.

M. chlorantha (greenish-yellow). — A newly-introduced species with erect, cylinder-shaped stems, 6 in. high, clothed with numerous tubercles, which are tipped with clusters of long, silvery, interlacing, hair-like spines, and a few stouter blackish ones. The flowers are described as greenish-yellow, so that they are not likely to add much to the beauty of the plant, which is recommended because of the attractiveness of its stem and spines. It is a native of Mexico and Texas, whence it was introduced some two years ago. It requires cool-house treatment, and should be kept free from dust, which disfigures the white spines.

M. cirrhifera (twisted). — Like *M. bicolor*, this species owes its frequent occurrence in gardens to the symmetry and neatly-chiselled form of its stems, and not to any attraction possessed by its flowers. It will thrive anywhere where the sun can shine upon it, if sheltered from severe cold and wet. In a cottage window it may be grown, and kept for many years, without losing health or, on the other hand, increasing much in size. Its usual height is about 5 in., by 4 in. in diameter. The tubercles are angular at the base, and bear tufts of yellowish spines on their pointed apices. The flowers are small, and bright rose-coloured, but only rarely produced on cultivated plants. Introduced from Mexico in 1835.

M. clava (club-shaped); Bot. Mag. 4358.—In the size of its stem, and the large, brightly-coloured flowers it bears, this species may be compared with some of the Echinocactuses. The stem is from 1 ft. to 1= ft. high, 4 in. wide at the base, narrowing slightly upwards; the tubercles are 1 in. long, and nearly as much through at the base, their shape that of little pyramids, and their tips bear each from eight to eleven stout, straight spines, pale brown, with a little wool at the base. The flowers are borne on the top of the stem, two or three of them together; the sepals are green and red, and the spreading petals are straw-coloured and glossy, their edges near the top being toothed. In the centre of the shallow cup formed by the petals, and which measures nearly 4 in. across, the orange-coloured stamens are clustered, in a kind of disk, through the middle of which the yellow stigma projects. It is a native of Mexico, at an altitude of 5000 ft. Introduced in 1848, when it flowered at Kew, in June, at which time it flowers almost every year now. A warm greenhouse affords the most suitable conditions for it; but, unless it is kept in full sunshine both summer and winter, and perfectly dry during the latter season, it will not produce any flowers. As a flowering plant, it ranks amongst the very best of the Mamillarias. It is easily propagated from seeds ripened on cultivated plants.

M. dasyacantha (thick-spined).—Stem 2 in. to 3 in. high, almost globular, and covered with spiral whorls of tiny tubercles, in the grooves of which is a little whitish wool, which falls away as the tubercles ripen. The spines upon the tubercles are arranged in little stars, with an erect central one. The flowers are small, and spring from the centre of the stem. This is one of the Thimble Cactuses, and is too small to have any great attractions, either in stem or flowers. It is, however, a pretty plant, especially when studded with its ruby-like flowers, which look like coloured Daisies growing upon a dense tuft of hairs. It is a native of Mexico, where it grows on high mountains among short grass and other herbage.

M. discolor (spines two-coloured).—Stem globose, about 4 in. in diameter; tubercles smooth, egg-shaped, their bases embedded in white wool, their tips crowned with stellate tufts of short, reddish spines. Flowers numerous, and borne from almost all parts of the stem, less than 1 in. wide, and composed of a single whorl of narrow, reflexed, rose-purple petals, surrounding a large, disk-like cluster of yellow stamens. The flowers are so short that they are half hidden by the tubercles. It is a native of Mexico, where it grows on rocks, in warm, sheltered places. Under cultivation it thrives when grown on a dry shelf in a warm house, and kept moist in summer, but perfectly dry in winter.

M. dolichocentra (long-spurred); Fig. 58.—Apparently this is a variable species; at all events, plants of widely different habit are found under this name, one of them represented in the Figure here, another in the *Garden,* Vol. XVII., whilst others are figured or described in other books. What is known at Kew as the true plant is that here figured. This has a stout stem, about 8 in. high and 3 in. wide, and covered with smooth cone-shaped mammae, with woolly bases and stellate tufts of spines on their tips. The flowers are produced about 1 in. from the top of the stem, and are less than 1 in. wide; they are, however, often very numerous, sometimes a closely-set ring of them surrounding the stem, like a daisy chain, their colour being pale purple. Below the flowers there is often a whorl of club-shaped fruits, > in. long, and rose-coloured. These contain numerous little black seeds, which, when ripe, may be sown in pots of very sandy loam. The plant is a native of Mexico, and flowers in summer. It thrives in a tropical temperature, and enjoys a daily syringing overhead on bright days in summer, but in winter requires little or no water.

FIG. 58.—MAMILLARIA DOLICHOCENTRA

M. echinata (hedgehog-like).—A charming little plant, with very small stems, clustered together in a cushion-like tuft, each stem less than 1 in. wide; but a well-grown specimen is composed of dozens of these, packed almost one on top of the other. The tubercles are hidden by the star-like spine clusters which cap them, and look like a swarm of insects. Flowers very small, rose-coloured, and lasting only about a day. These are succeeded by numerous currant-like red berries, so numerous, in fact, that the plants look as if thickly studded all over with coral beads. The central stem is sometimes about 6 in. high, those surrounding it being shorter and shorter, till the outside ones rise only just above the soil. A well-grown plant of this is strikingly pretty, even when not in fruit. It is a native of Mexico, and requires the treatment of a warm house. A few pieces of broken brick should be placed upon the surface of the soil about the base of the plant, as the stems like to press against, or grow upon, anything in the nature of rocky ground.

M. echinus (hedgehog-like); Fig. 59.—A distinct and pretty little plant, the largest specimen having a stem about the size and shape of a small hen's-egg, completely hidden under the densely interwoven radial spines, which crown the thirteen spiral rows of tubercles, and are almost white when mature. The tubercles are = in. long, and, in addition to these white radiating spines, they also bear each a stout spike-like spine, growing from the centre of the others. This spine gives the plant an appearance quite distinct from all other cultivated Mamillarias. The flowers are produced two or three together, on the top of the stem, and they are nearly 2 in. long, cup-shaped, and coloured yellow; they usually appear about June. As yet this species is rare in cultivated collections. It comes from Mexico, where it is found growing on limestone hills, in hot and arid localities. Under cultivation it requires a warm greenhouse temperature, exposure to bright sunshine all the year round, with a moderate supply of water in summer, and none at all during winter. A few large pieces of broken brick or sandstone placed in the soil, just under the base of the stem, afford the roots conditions suitable to their healthy growth.

FIG. 59.—MAMILLARIA ECHINUS

M. elegans (elegant).—A small species, grown only for the prettiness of its stem, flowers rarely, if ever, being borne by it under cultivation. The stem is 2 in. high and wide, globose, with small conical tubercles, which, when young, are woolly at the tips. Spines short and slender, about twenty, arranged in a star on each tubercle, with four central ones a little longer than those which surround them; the colour of the spines is whitish, with brown tips. Native country Mexico, on high exposed hills; in this country it requires greenhouse treatment. Introduced about 1850.

M. elephantidens (elephant's-tooth); Fig. 60.—One of the largest and most remarkable of all garden Mamillarias. Stem globose, depressed, 6 in. to 8 in. in diameter, and bright shining green. Tubercles smooth, round, 1= in. long, furrowed across the top, which is at first filled with wool, but when old is naked. At the base of the tubercles there is a dense tuft of white wool, and springing from the furrows are eight radiating recurved spines, and three short central ones, all strong, stiff, and ivory-white, tipped with brown. The flowers are 3 in. wide, and are composed of a circle of violet-coloured sepals, with white margins, and a second circle of petals which are bright rose, pale purple at the base, a line of the same colour extending all down the middle. The stamens are numerous, with long purple filaments and yellow anthers, and the pistil is stout, erect, projecting above the stamens, with a radiating stigma. Flowers in autumn; native country, Paraguay. Under cultivation, it grows quicker than is usual with plants of this genus, and it is also exceptional in the regular and abundant production of its flowers. It has been a rarity in European collections for many years, and, although easily grown, it is often killed through wrong treatment. A cool greenhouse or sunny frame in summer, plenty of water whilst growth is active, and a light, well-drained soil, suit it best; whilst during winter it must be kept perfectly dry, and protected only from frost. In a tropical house, it is invariably sickly, and altogether unsatisfactory.

FIG. 60.—MAMILLARIA ELIPHANTIDENS

M. elongata (elongated).—A small, cushion-like kind, with the stems in tufts, owing to their producing offsets freely from the base, the tallest of them being about as high and as thick as a man's thumb. The tubercles are short, crowded, and hidden under the star-clusters of reddish-yellow spines. There are no central spines in this kind. The flowers are produced in the axils of the tubercles from all parts of the stem, a large tuft of stems being thickly studded with circles of tawny yellow petals, which are only about = in. long. The berries are bright coral-red, and about the size of a date stone. There are several varieties of this species, under the names of *intertexta, rufescens, rutila, subcrocea,* and *supertexta.* These differ only slightly either in the length or thickness of the stems or in the colour of the spines. All of them may be grown in a cold frame, or in a window where the sun can shine upon them; or they may be grown along with tropical kinds. For small cases in windows, these little Thimble Cactuses are amongst the most suitable. They are natives of high mountains in Mexico, and have been cultivated in Europe over forty years.

M. fissurata (fissured); Fig. 61.—In appearance, this rare species mimics some of the Gasterias, and is so different from all the kinds hitherto described, that very few people unacquainted with it would suspect that it belonged to the same genus as *M. elongata* or *M. dolichocentra*. Indeed, some botanists have made a separate genus of this and several other plants of the same peculiar appearance, calling them *Anhalonium*. *M. fissurata* is like a whip-top in shape, the root being thick and woody, and the tubercles arranged in a thick layer, spreading from the centre, rosette-like. A living plant in the Kew collection is 2 in. high by 4 in. wide, the tubercles being triangular in shape, = in. thick, wrinkled, with an irregular furrow on the upper surface. The flowers grow from the middle of the stem, and are 1= in. wide, and rose-coloured. Native of Mexico, on hard gravel or limestone soils. We know of no plant in English collections, except that at Kew, which was introduced from Mexico in 1886. It flowers in September and October.

FIG. 61.—MAMILLARIA FISSURATA

M. floribunda (free-flowering).—A French writer on Cactuses, M. Labouret, calls this a species of Echinocactus, but it resembles so closely another species included by him in Mamillaria, viz., *M. atrata*, that we see no good reason for separating the two into different genera. *M. floribunda* has an irregular conical stem, about 5 in. high by 4 in. wide at the base, round nut-like tubercles the size of filberts, crowned with star-tufts of spines > in. long, stiff, and brown, about ten spines being set with their bases in a small disc-like pad of dirty-white wool. The flowers are very numerous, covering the whole of the stem-top, from which they stand erect, so as to form a dense bouquet of rose-coloured petals. Each flower is 2 in. long. Native of Chili; introduced about 1835. Flowers in summer. This handsome kind will thrive in a window, and, if well supplied with fresh air, sunshine, and sufficient water to keep the soil moist, it will flower almost every year. It must have no water in winter.

M. gracilis (slender).—A small Thimble Cactus, remarkable for its proliferous stems, a single stem 2 in. high producing all round its upper half numerous, offshoots, which fall to the ground and grow. In this way a tuft of stems is soon developed round the first one. If these offshoots are removed as they appear, the stem will grow longer and stouter than it does when they are left. Tubercles small, green, crowded; spines in a stellate tuft, short, curved, pale yellow or white. Flowers as in *M. elongata*, to which this species is closely allied. In window cases, or on a shelf in a cool greenhouse, it will grow and multiply rapidly. Like the bulk of the caespitose, or Thimble Cactuses, it does not make much show when in flower; and it is only its stems, with their white stars of spines and clusters of little offsets hanging about them, that are attractive. Native of Mexico; introduced about 1850. There is a variety known as *pulchella*, in which the spines are of a yellow hue.

M. Grahami (Graham's).—A pretty little species, with globose stems, scarcely 3 in. high, and nearly the same in diameter, branching sometimes when old; tubercles < in. long, egg-shaped, corky when old, and persistent. Spines in tufts of about twenty, all radiating except one in the centre, which is hooked; they are about = in. long. Flowers 1 in. long, usually produced in a circle round the stem. Fruit a small, oval berry, = in. long. This is a native of Colorado, in mountainous regions. It is very rare in cultivation. The flowers are developed in June and July.

M. Haageana (Haage's); Fig. 62.—The habit of this is shown in the Figure, which is reduced to about one-fourth the natural size. As the stem gets older, it becomes more elongated. Tubercles small, four-sided at the base, pointed at the top, where the spines are arranged in a star, about twenty of them on each tubercle, with two central ones, which are longer, stiffer, and much darker in colour than those on the outside; flowers small, almost hidden beneath the spines, bright carmine-rose; they are produced on the sides of the upper portion of the stem in June. There is a close resemblance between this and *M. cirrhifera*, and the treatment for both should be the same. Mexico, 1835.

FIG. 62.—MAMILLARIA HAAGEANA

M. longimamma (long-tubercled); Fig. 63.—A well-marked species in the size of its mammae, or tubercles, which are at least 1 in. long by 1/3 in. in diameter, terete, slightly curved, and narrowed to a pointed apex, the texture being very soft and watery. Each tubercle bears a radiating tuft of about twelve spines, one central and projecting outwards; they are pale brown when old, and white when young; their length is about = in. A tuft of short, white wool is developed at the base of the spines on the young mammae. The stem is seldom more than 4 in. in height, and it branches at the base when old. Flowers large and handsome, citron-yellow; the tube short, and hidden in the mammae; the petals 1= in. long, narrow, pointed, and all directed upwards; stamens numerous, short. Flowering season, early summer. Native country, Mexico. It requires greenhouse treatment, or it may be placed in a sunny frame out of doors during summer. It is not easily multiplied from seeds, but is free in the production of offsets from the base of the stem.

FIG. 63.—MAMILLARIA LONGIMAMMA

M. macromeris (large-flowered); Fig. 64.—Stem about 4 in. high, naked at the base, woody and wrinkled when old. Tubercles as in *M. longimamma*, but with curving radial spines, like needles, often 2 in. in length, white or rose-tinted when young, almost black when old. Flowers from the centre of the stem, 3 in. long, and about the same in width; the petals regular and spreading, as in the Ox-eye daisy; stamens numerous, short, forming a disk; colour carmine, almost purple just before fading. Flowering season, August. Native of Mexico, where it is found in loose, sand on hillocks, generally about the roots of Acacias. It is one of the most beautiful of all Mamillarias; but it is, as yet, rare in collections. It requires the same treatment as *M. longimamma*, except that, owing to the woody nature of its rootstock, and its long, tap-like roots, it should be planted in pans instead of pots, using a compost of rough loam, mixed with lumps of broken brick or limestone.

FIG. 64. — MAMILLARIA MACROMERIS

M. macrothele (large-nippled); Bot. Mag. 3634, as *M. Lehmanni*. — This belongs to the same group as *M. cirrhifera*, but is distinguished by its large mammae, which are four-angled at the base, > in. long, narrowed to a point, upon which is a tuft of wool and a cluster of about eight spines, = in. long, spreading, reddish-brown in colour, the central one being almost black, 1 in. long, and pointing downwards. In the axils of the mammae are tufts of white wool. Flowers on the top of the stern, erect, spreading, about 1= in. across when expanded; the petals overlapping, and pale yellow; the stamens red at the base, arranged in a dense cluster, and the rays of the stigma spreading over them. Flowering time, early summer. Native country, Mexico, on prairie lands, at high elevations. This species is almost hardy in the warmer parts of this country, suffering from damp rather than frost in winter. The stem is not particularly handsome, but the flowers are large and bright, and they are produced annually by plants which are grown in a cool, well-aired greenhouse or frame, with the sun shining on them all day.

M. micromeris (small-flowered); Fig. 65.—A small, cushion-like plant, with a stem never more than 1= in. across by about 1 in. in height, so that it has the appearance of a small, flattened ball, with a raised, disk-like portion on the top. The mammae are very small, and they are completely hidden by the numerous fine, white, silky spines and wool which spring in tufts from the apex of each mamma, and interlace so as to form a spider-web-like net all over the stem. The flowers are small, and they spring from the centre of the disk-like top of the stern; they are composed of from three to five sepals, and five petals, which are whitish or pink, and measure about < in. across when open. Native country, Mexico, where it is found only in naked places on mountain tops or sides where limestone is plentiful. It requires much care under cultivation, water in excess being fatal to it, and a soil of the wrong sort soon killing all its roots. It is cultivated at Kew in a small pot, in a mixture of loam and lime rubbish, and grown in a warm greenhouse.

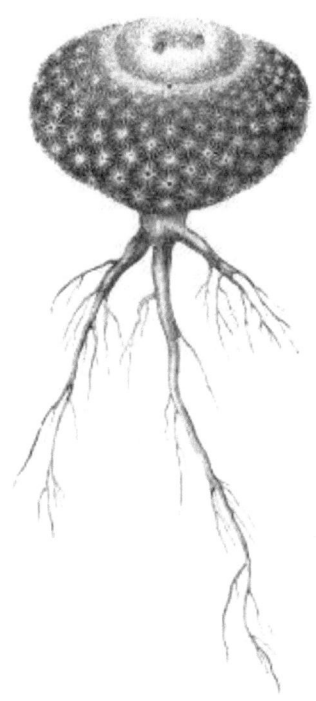

FIG. 65.—MAMILLARIA MICROMERIS

M. multiceps (many-branched).—Stem short, with numerous branches, which again push forth other branches, so that a dense tuft of dumpy, globose stems is formed. The mammae are small, and arranged closely together, and they each bear a tuft of whitish wool, with a radiating cluster of spines, which are soft, almost hair-like in texture, the inner ones being stiffer, and coloured dull yellow. The flowers are small, and almost hidden by the spines and tubercles; they are pale yellow, with a line of red down the middle of each petal. Native country, Mexico. This plant should be grown on a shelf in a cool greenhouse—anything like a stove temperature being fatal to it. As a flowering plant it is of no value, but the compact tuft formed by its numerous stems, with their attractive spines, renders it worthy of cultivation.

M. Neumanniana (Neumann's).—This is a member of the group with angular tubercles and comparatively small flowers. It has a stem about 6 in. high, cylindrical, the tubercles arranged spirally, their bases compressed, four or five-angled, and with a tuft of white wool in their axils. The areoles or tufts on the tops of the mammae are large, and the spines are about seven in number, = in. long, and of a tawny-yellow colour. The flowers are produced near the top of the stem; they are about = in. long, and rose-red in colour. Native country, Mexico. It requires the same treatment as *M. cirrhifera*.

M. Ottonis (Ottoni's); Fig. 66.—A very distinct and pretty plant is cultivated under the name at Kew; but there are, apparently, two different species under the same name—the one being spiny and large in the stem; the other, which is here shown, having a small, compressed stem, 3 in. across, numerous compressed tubercles, and short, hair-like spines. The flowers, which are large for the size of the plant, are white, and are developed in May and June. Native country, Mexico; introduced in 1834. It requires similar treatment to *M. micromeris*.

FIG. 66.—MAMILLARIA OTTONIS

M. pectinata (comb-like); Fig. 67.—Stems globose, from 2 in. to 3 in. in diameter; the rootstock woody; the tubercles arranged in about thirteen spiral rows, swollen at the base, and bearing each a star-like tuft of about twenty-four stiff, brown, radial spines, without a central one; the length varies from = in. to 1 in., and they are comb-like in their regular arrangement. When not in flower, this species bears a close resemblance to small plants of *Cereus pectinatus*. Flowers terminal, solitary, large, their width quite 3 in. when fully expanded; sepals reddish-green; petals rich sulphur-yellow; filaments reddish, very numerous; the flowers open at noon, and close after about two hours, even although the sun be shining full upon them. Flowering season, June to August. Native country, Mexico, on slopes of limestone hills. Although long since known to botanists, this pretty species has only lately found its way into English gardens. It is attractive even when not in flower. It requires warm greenhouse treatment, with exposure to full sunshine; during late autumn it should have plenty of air to ripen the new growth made whilst flowering. In winter it should have a dry position near the glass.

FIG. 67.—MAMILLARIA PECTINATA

M. phellosperma (corky-seeded).—A pretty plant, resembling *M. Grahami* in all points except the seed, which, as is denoted by the name, is half enveloped in a corky covering, suggesting acorns. Stems simple, sometimes proliferous at the base, globose when young, afterwards almost cylinder or pear-shaped, 5 in. high, 2 in. in diameter; tubercles = in. long, arranged in twelve spiral rows, slightly woolly in axils. Spines radiating, in two rows, about fifty on each tubercle, the three or four central ones being hooked at the tips or sometimes straight; length, = in. to 1= in. Flowers (only seen in the dried state) 1 in. long and wide. Native of the dry gravelly hills and sand ridges in California and Colorado, and, therefore, requiring greenhouse treatment. This plant is cultivated in the Kew collection, but it has not been known to flower there. It is one of the most ornamental of the very spinous species, the radial spines being almost white, whilst the central ones are black, and look like tiny fishhooks. A large proportion of these Mamillarias are far more interesting in the form and arrangement of their tubercles and spines than in any floral character, and it is on this account that so many which are insignificant as flowering plants are included here.

M. pulchra (handsome).—Stem globose when young and cylindrical when old, flattened at the top; height from 4 in. to 6 in.; tubercles large, egg-shaped, arranged in from eleven to thirteen spiral rows; spines in compact tufts, their bases set in whitish wool, irregular in length, and almost covering the whole of the stem. Flowers medium in size, developed near the top of the stem from the woolly axils of the tubercles; colour bright rose. Native of Mexico. Flowering season, June. Introduced in 1826. A rare kind nowadays, though one of the prettiest. It should always be grown in a warm house. It has been also called *M. pulcherrima*.

M. pusilla (small).—A tiny tufted plant, belonging to the group known as Thimble Cactuses. It has stems 2 in. high; short, dark green tubercles, with tufts of whitish wool in the axils; spines thin and bristle-like, twisted, nearly 1 in. long, almost hiding the stem; they are whitish, with black tips. The flowers are yellowish-white, with streaks of red. Common in Mexico. Flowering season, May. It should be grown in a frame in summer, and wintered on a shelf in a warm greenhouse. It would, no doubt, thrive in a window if kept in a sunny position and placed under a glass shade. A variety known as *texana* differs in being more densely clothed with spines. We have seen it grown into large clumps, covering a space 1 ft. in diameter, with dozens of erect little pyramids of whitish spines.

M. pycnacantha (densely spined); Bot. Mag. 3972.—The name for this kind is rather misleading, the spines being both fewer and less conspicuous than in many other species of Mamillaria. Stem about 6 in. high, nearly globose; tubercles—rather large, swollen, with tufts of short white wool in their axils, and stellate clusters of spines springing from disks of white wool on the top. The spines are = in. long, slightly recurved, flattened, and pale brown. Flowers large, clustered on the top of the stem, about half a dozen opening together; width 2 in.; petals numerous, narrow, toothed at the tips, spreading; colour a deep sulphur-yellow, anthers orange. Native of Oaxaca, Mexico. Flowering season, July. Introduced 1840. This is a beautiful flowering plant, more like an Echinocactus than a Mamillaria.

It should be grown in a warm greenhouse all the year round. Old stems develop offsets from the base, by which the species may be multiplied.

M. sanguinea (bloody); Fig. 68.—This is closely related to *M. bicolor*, but differs in having an unbranched stem and numerous richly-coloured flowers. The stem is stout, 6 in. high, and 4 in. through; tubercles crowded, short, bearing stellate tufts of shortish spines, and projecting longer ones, all being bristly and pale yellow, except those on the youngest tubercles, which are golden. The flowers are borne in a crowded circle on the top of the stem, just outside the cluster of young yellow spines, a strong plant having about forty flowers open together. Each flower is about = in. long and wide, and coloured bright crimson, with yellow anthers. Native of Mexico. Flowers in June. It should be grown along with *M. bicolor*. The plant figured is a young one, showing the spines much longer than is usual on mature specimens.

FIG. 68.—MAMILLARIA SANGUINEA

M. Scheerii (Scheer's). — Stem 7 in. high, and 5 in. in diameter at the base; tubercles large, swollen, somewhat flattened, pale green, watery, woolly in the axils, the tops crowned with about a dozen brown spines, 1 in. long, one central, the others radial. Flowers terminal, erect, with several whorls of spreading, recurved petals, the lower ones tinged with crimson, the upper pale yellow, and forming a shallow cup, 2 in. across; anthers forming a compact sheaf in the centre. Flowers in summer. This distinct and very pretty species was introduced many years ago from Mexico, where it was discovered in 1845 by a Mr. Potts, to whose love for these plants we are indebted for a great many choice kinds collected and sent to England by him. It grows naturally in a red, sandy loam, and under cultivation requires warm-house treatment, except during the autumn, when it may be placed in a frame and exposed to full sunshine and plenty of air.

M. Schelhasii (Schelhas'). — A pretty little tufted kind, its habit and size being shown in Fig. 69. The stem produces offsets freely at the base, which grow into full-sized stems, and develop young ones, till a compact cushion is formed. Tubercles closely arranged, cylindrical, shining green, with fifteen to twenty radial, white, hair-like spines, = in. long, and three inner ones, which are thicker, purplish in colour, usually only one being hooked. Flowers white, with a line of rose down the middle of each petal, > in. across. Flowering season, beginning of summer. Native of Mexico. It may be grown out of doors in a sunny position in summer, and wintered on a shelf in a greenhouse.

FIG. 69.—MAMILLARIA SCHELHASII

M. Schiedeana (Schiede's). —Stem globose, 3 in. to 5 in. high, thickly clothed with long, narrow, pointed tubercles, the bases of which are set in white wool, whilst the apices are crowned with tiny stars of white silky spines; more like the pappus of a Composite than the spines usually found on Cactuses. A healthy plant has a very pretty and silky appearance which cannot well be described. The flowers are small and unattractive; they are succeeded by the red fruits, which remain on the plant a long time, and add to its beauty. Native country, Mexico. Introduced 1838. Should be grown in a greenhouse where frost is excluded, and where there is plenty of sunlight at all times. It is easily increased, either from seeds or by means of the offsets developed at the base of old stems.

M. semperviva (ever-living); Fig. 70. —Stem pear-shaped, 3 in. wide, the top slightly depressed. Tubercles conical, < in. long, their bases set in a cushion of white wool, their tips bearing tiny tufts of wool, and four small spines, which fall away on the tubercles becoming ripe, leaving two short, diverging, central spines. Flowers small, not ornamental, and scantily developed near the outside of the top. Native of Mexico; in meadows and thickets near Zimapan, at 5000 ft. elevation. It thrives with us when grown in a frame in summer, and wintered in a cool greenhouse or frame.

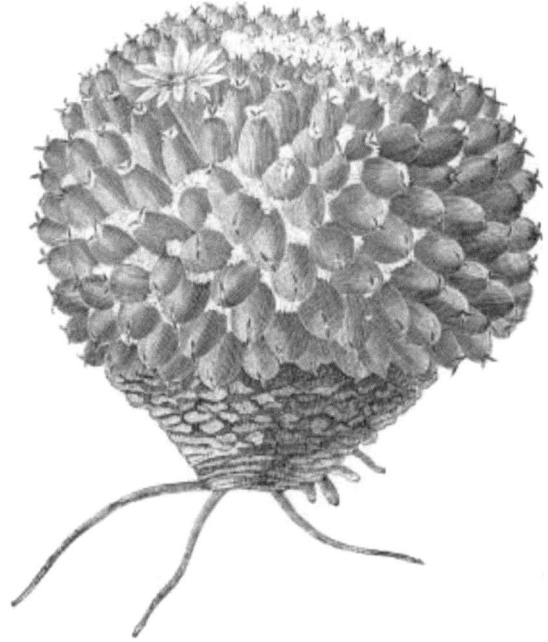

FIG. 70.—MAMILLARIA SEMPERVIVA

M. senilis (hoary).—Stem about 3 in. high, spherical, unbranched, except when very old, when it becomes proliferous at the base; tubercles crowded, small, arranged spirally, and crowned with clusters of long, radiating spines, which are almost white, hair-like, and become thickly interwoven, as in the Old Man Cactus *(Pilocereus senilis)*. The central spine is black, and hooked at the tip. Flowers on the top of the stem, near the centre; the petals toothed, spreading, and forming a deep cup, with a cluster of tall stamens standing erect in the middle; colour bright scarlet. The flowers, which appear in summer, remain open about eight hours. Native country unknown;

cultivated in France in 1845. This plant is difficult to preserve in health, the best method being that of grafting it on to a short Cereus, or a robust kind of Mamillaria, such as *M. cirrhifera*. It is a pretty plant at all times, even when dead, for we have seen plants of it preserve the appearance of live specimens long after they have rotted and dried up in the centre, nothing remaining but the shell formed by the skin and silvery spines. There is a close resemblance between this species and *M. Grahami*.

M. stella-aurata (golden star).—This little plant obtains its name from the rich golden-yellow of its stellate clusters of spines, which are arranged thickly on the tips of the small, pointed tubercles. It belongs to the group called Thimble Cactuses, of which it is one of the prettiest. The stems are tufted, branching freely at the base, and rising to a height of about 2 in. Flowers small, whitish, and much less ornamental than the berry-like fruits which succeed them, and which are egg-shaped, = in. long, and a deep rose-colour. *M. tenuis* is a variety of this, with almost white spines. Native of Mexico. Introduced 1835. May be cultivated under a bell-glass in a room window, the only danger being damp during winter, which must be carefully avoided.

M. sub-polyhedra (usually many-sided); Fig. 71.—Stem simple till it becomes old, when it develops offsets at the base, broadly cylindrical, 8 in. high, 5 in. in diameter. Tubercles four-sided at base, prism-shaped, bearing pads of white wool in the corners at the base, and crowned with tufts of from four to seven spines, usually all radial, sometimes one central. The flowers, which usually appear in May, are arranged in a zone on the top of the old stems; sepals greenish-yellow, petals bright red. Fruit 1 in. long, pear-shaped, scarlet. Native of South Mexico, at high elevations. It may be grown outside in summer, and wintered in a heated greenhouse or frame. This is a singular-looking plant, the tubercles having an appearance suggestive of carving. It is a slow grower, and requires careful attention in winter, when sometimes the roots all perish and the base of the stem rots.

FIG. 71.—MAMILLARIA SUB-POLYHEDRA

M. sulcolanata (woolly-grooved); Fig. 72.—Stem simple when young, proliferous at the sides when old, the young plants developing from the apices of the tubercles, and not in the axils, as is usual. The tubercles are nut-shaped, large, the bases surrounded by white wool, the points bearing eight to ten rigid, brown spines, all radiating from a little pad of wool. Flowers large, nearly 2 in. across, bright yellow, poppy-scented, the spread of the petals suggesting Paris Daisies; they are freely developed on the apex of the stem in June, and on till August. Fruit egg-shaped, glaucous-green. Native country, South Mexico; introduced 1836. This charming little plant should be grown in a frame exposed to full sunshine all summer, and removed to a shelf in a warm greenhouse in winter. With such treatment it grows and flowers freely. Grafted on to a Cereus or Opuntia it is healthier than when on its own roots.

FIG. 72.—MAMILLARIA SULCOLANATA

M. tetracantha (four-spined); Bot. Mag. 4060.—Stem the size and shape of an ostrich's egg, thickly studded with small, conical tubercles, woolly at the base, the apices bearing each four spreading spines, > long, rather stout, straight, brown when young, becoming almost white with age. Flowers numerous, small, arranged as in *M. sanguinea*, to which and *M. cirrhifera* this species is closely related. They are bright rose in colour, with orange-yellow anthers, and are developed in July. Native of Mexico. Requires the same treatment as *M. cirrhifera*.

M. tuberculosa (tubercled).—This is a very pretty and distinct plant, of recent introduction, and easily cultivated. It has a central stem, 6 in. high by 2 in. in diameter, conical in shape, and surrounded at the base by globose branches or offsets. The tubercles are closely set in numerous spiral rows, and are = in. long, rather narrow, pointed, with a crown of radial spines, very slender, hair-like, white, and = in. long; central spines three or four, = in. long. At the base of each tubercle is a pea-like tuft of white wool. In this kind the spines fall from the old tubercles, which are persistent, gradually hardening to a cork-like substance. The flowers are produced in the apex of the stem, and are 1 in. long and wide, daisy-like, pale purple in colour; they are succeeded by red, oval berries, which are as pretty as the flowers. About five flowers are developed on each stem annually—May and June. Native of Mexico, in the mountains. It thrives when grown in an ordinary greenhouse, on a shelf, in full sunshine.

M. turbinata (top-shaped); Bot. Mag. 3984 .—Stem globose, depressed at top, about 3 in. in diameter, pale glaucous-green; tubercles quadrangular, flattened at the apex, and bearing, when young, from three to five erect, slender, hair-like spines, which fall off soon after the tubercles ripen, exposing little depressions or umbilica, and giving the stem a bald, pudding-like appearance, quite distinct from any other kind. Flowers from the centre of the stem, short, about 1 in. across, pale yellow, with a reddish tint outside; anthers yellow. Two or three flowers are usually expanded together in the month of June. Native country, Mexico.

M. uncinata (hooked).—Stem globose, simple, about 4 in. in diameter; tubercles closely pressed against each other at the base, where they are four-angled; in length they are < in., and they are blue-green in colour. Apex bearing four short spines, arranged crosswise, and < in. long; central spine slightly longer, yellow, and hooked. The flowers are 1 in. long and wide, erect, the tube hidden by the young mammae, amongst which they appear in May and June; they are purple in colour, a line of deeper tint running down the middle of each petal. Like all the kinds with short, angular tubercles, this species is easily managed, flowers freely and profusely, and always ripens seeds. Native of Mexico. It may be grown in a frame, or even out of doors, all through the summer, removing it to a greenhouse for the winter.

M. vetula (old).—One of the small Thimble Cactuses, its stems seldom exceeding 3 in. in height by 1= in. in diameter. Tubercles < in. long, conical, with a radial crown of fine, hair-like yellow spines, < in. long, and a solitary central spine, = in. in length, and coloured red. Flowers terminal, just peeping above the tubercles; sepals and petals acute, yellow, > in. long; anthers yellow; stigma white. An old garden plant, introduced from Mexico. It flowers in May and June. For its cultivation it may be treated as recommended for *M. pusilla*.

M. villifera (hair-bearing).—Stem similar to the last, but usually proliferous at the base; tubercles angular, short, woolly in the axils, and bearing four rigid, short, reddish-brown spines on the apex. Flowers pale rose, with a line of purple down the middle of each petal; they are developed near the top of the stem, in May. Native country, Mexico. This plant thrives if treated as recommended for *M. pusilla*. There are several varieties known, distinguished by their paler or darker flowers, or by a difference in the length and arrangement of the spines.

M. viridis (green).—Stem 4 in. high by 3 in. in diameter, proliferous at the base; tubercles short, four-angled, crowded in spiral rows, woolly at the base, bearing each five or six radiating hair-like spines on the apex, and one central erect one, none more than < in. long. Flowers erect, on top of stem, with recurved, pale yellow petals, 1 in., long; they are produced in May and June. Introduced from Mexico in 1850. It may be grown in a sunny frame out of doors during summer, and on a dry, warm greenhouse shelf in winter.

M. vivipara (stem—sprouting).—A tufted, free-growing Thimble Cactus, producing its small stems in such profusion as to form a cluster as much as 3 ft. in diameter. The small tubercles are hidden by the numerous radial spines, which are in clusters of about twenty; they are white, hair-like, stiff and = in. long; the central spines, numbering from four to six, are a little longer. Flowers from apex of stem, 1= in. long and wide, and composed of about thirty fimbriated sepals and twenty-five to forty narrow petals; colour bright purple. Fruit = in. long, pale green when ripe. The flowers, which appear in May and June, usually expand after mid-day. Native of Louisiana. In the North-West plains and Rocky Mountains of North America this plant is abundant, often forming wide cushion-like tufts, which, when covered with numerous purple, star-like flowers, have a pretty effect. In Utah and New York it is commonly cultivated as a hardy garden plant, bearing exposure to keen frosts and snow without suffering; but it would not thrive out of doors in winter with us, unless covered by a handlight during severe weather, and protected from heavy rains in winter. It likes a strong, clayey soil.

M. v. radiosa (Fig. 73). — This variety is distinguished by its larger flowers and shorter spines.

FIG. 73.—MAMILLARIA VIVIPARA RADIOSA

M. Wildiana (Wild's).—An old garden Cactus, and one of the prettiest of the tufted, small-stemmed kinds. Its largest stems are 3 in. high by about 1= in. in diameter, and bear spiral rows of clavate, dark green, crystallised tubercles, = in. long, with about ten radial white spines, = in. long, the three upper spines, together with the solitary central hooked one, being yellow. Flowers small, numerous on the apices of the stems, rose-coloured, lined with purple; they are developed in summer. This also forms dense tufts of stems. A specimen at Kew, only a few years old, has already over thirty heads. It is a native of Mexico, at an altitude of 5000 ft., growing on lava and basalt, and even on the trunks of trees. For its cultivation, a shelf in a sunny greenhouse is a most suitable position, both in winter and summer. Introduced 1835.

M. Wrightii (Wright's).—This is a charming little plant, of something the same character as *M. dolichocentra*. It has not long been cultivated in gardens, but being easy to manage, and exceptionally pretty, it is sure to become a favourite as it gets known. Stem rounded above, narrowed and peg-top-like at the base, the top flattened, about 3 in. across, height about the same. Tubercles conical, = in. long, shining green, and bearing a tuft of six or eight spines, which are straight, hair-like, white, and = in. long; there are two central spines, of same length, and hooked. Flowers in the top of the stem, 1 in. long and wide, bright purple; they are succeeded by egg-shaped, purple berries, 1 in. long, and prettily arranged among the tubercles. In England a warm house seems most suitable for this species. It likes plenty of moisture and sunlight during the summer, whilst making new growth; but in winter, when at rest, it ought to be kept on a shelf, and just moistened overhead in bright weather. There are healthy examples of it at Kew. Flowering season, May and June. Native country, Mexico. Introduced about 1878.

M. Zuccchariniana (Zuccharini's).—Stem simple, globose, often attaining a height of 10 in. by about 7 in. in diameter. Tubercles dark green, conical, 1/3 in. long, = in. broad at base, naked at the point, but with four to six spines springing from the areole a little below the point; spines ash-coloured, stiff, black-tipped. Flowers in a ring about the top of the stem, length 1 in., the tube enveloped in long, black, twisted hairs; sepals brown-purple; petals narrow, sharp-pointed, purple-rose coloured; stamens white and yellow; stigma rose-coloured. Flowers in June and July. Native of Mexico. A large, handsome-stemmed kind, easily kept in health, and flowering freely if grown on a shelf in a cool greenhouse in winter, and placed in a warm, sunny position out of doors in summer. It produces seeds freely, and pretty plants, 3 in. or more in diameter, may be obtained in two years from seeds. By grafting it, when young, on the stem of a Cereus or cylindrical Opuntia, a healthy, drumstick-like plant is easily obtained.

CHAPTER XIII.

THE GENUS LEUCHTENBERGIA.

(Named in honour of Prince Leuchtenberg.)

AMONG the many instances of plant mimicry that occur in the Cactus order, the most remarkable is the plant here figured. Remove the flower from Leuchtenbergia, and very few people indeed would think of calling it a Cactus, but would probably consider it a short-leaved Yucca. In habit, in form, in leaf, and in texture, it more resembles a Yucca or an Agave than anything else, and when first introduced it was considered such by the Kew authorities until it flowered. The leaves, or rather tubercles, are sometimes longer and slenderer than in Fig. 74. The nearest approach to this plant is *Mamillaria longimamma*, in which the tubercles are 1 in. or more long, finger-shaped, and crowned with a few hair-like spines. But the Leuchtenbergia bears its flowers on the ends of the tubercles, and not from the axils, as in all others. This peculiarity leads one to infer that tubercles are modified branches, the spines representing the leaves. Some species of Mamillaria and Echinocactus develop young plants from the tops of their tubercles; and this also points to the probability that the latter are branches. In Leuchtenbergia, the tubercles fall away as the plant increases in height, leaving a bare, woody stem similar to that of a Yucca.

Cultivation.—The Leuchtenbergia has always been difficult to keep in health. It thrives best when kept in a warm, sunny house during winter, and in an exposed, airy, warm position under a frame during summer. It may be watered regularly whilst growing—that is, from April to September—and kept quite dry all winter. The soil should be well-drained loam, and the roots should have plenty of room. A specimen may be seen in the Kew collection.

Propagation.—This may be effected from seeds, or by removing the head from an old plant, putting the former in sand, and placing it under a bell-glass to root, watering it only about once a week till roots are formed. The old stem should be kept dry for about two months, and then watered and placed in a sunny, moist position, where it can be syringed once a day. A shelf in a stove is the best position for it. Here it will form young buds in the axils of the withered tubercles, and on the edges of the persistent parts of the tubercles themselves. They first appear in the form of tiny tufts of yellowish down, and gradually develop till the first leaf-like tubercle appears. When large enough, the buds may be removed and planted in small pots to root. If an old plant is dealt with in this way in April, a batch of young ones should be developed and rooted by October. Grafting does not appear to have ever been tried for this plant. When sick, the plant should be carefully washed, and all decayed parts cut away; it may then be planted in very sandy loam, and kept under a bell-glass till rooted.

FIG. 74. – LEUCHTENBERGIA PRINCIPIS

SPECIES.

L. principis (noble); Fig. 74.—This, the only species known, was introduced from Mexico to Kew in 1847, and flowered the following year. The plant attains a height of 1 ft. or more, the stem being erect, stout, clothed with the persistent, scale-like bases of the old, fallen-away tubercles, the bases having dried up and tightened round the stem. The upper part is clothed with the curved, leaf-like tubercles, from 3 in. to 6 in. long, grey-green in colour, succulent, with a tough skin, triangular, and gradually narrowed to a blunt point, upon which are half a dozen or more thin, flexuous, horny filaments, neither spines nor hairs in appearance, but almost hay-like; the central one is about 5 in. long, and the others about half that length. The flowers are borne on the ends of the young, partly-developed tubercles, near the centre of the head; they are erect, tubular, 3 in. to 4 in. long, scaly, gradually widening upwards; the sepals and petals are numerous, and form a beautiful flower of the ordinary Cactus type, quite 4 in. across, and of a rich, clear yellow colour. The anthers, which also are yellow, form a column in the centre, through which the nine-rayed stigma protrudes. Strong plants sometimes produce two flowers together.

CHAPTER XIV.

THE GENUS PELECYPHORA.

(From *pelekyphoros*, hatchet-bearing; referring to the shape of the tubercles.)

IKE Leuchtenbergia, this genus is monotypic, and it is also rare, difficult to cultivate, and exceptionally interesting in structure. It is closely related to the Mamillarias, as may be seen, by comparing the Figure here given with some of them; indeed, it was once known as *M. asellifera*, having been described under that name when first introduced, in 1843. From Mamillaria, however, it differs in the form of its tubercles, which are hatchet-shaped, and cleft at the apex, where each division is clothed with small, horny, overlapping scales, not unlike the back of a woodlouse—hence the specific name.

Cultivation.—The Hatchet Cactus grows very slowly, specimens such as that represented in our Illustration being many years old. We have seen healthy plants, freshly imported, grow for a few months, and then suddenly die, the inside of the stem rotting whilst outside it looked perfectly healthy. It is always grown on its own roots, but probably it would thrive better if grafted on the stem of some dwarf Cereus or Echinocactus.

FIG. 75.—PELECYPHORA ASELLIFORMIS

Propagation.—The propagation of Pelecyphora is easiest effected by means of seeds, which, however, are not always procurable. It is stated by Labouret, a French writer on Cactuses, that the first plants introduced arrived dead, but a few seeds were found in a withered fruit on one of the dead stems, and from these the first plants grown in Europe were raised. M. de Smet of Ghent, had a large stock of this Cactus a few years ago, and a German nurseryman, H. Hildmann, of Oranienberg, near Berlin, usually has many young plants of it for sale.

SPECIES.

P. aselliformis (woodlouse-like); Fig. 75.—The size, habit, and structure of this plant are so well represented in the Figure that little description is necessary. The stems are simple till they get about 3 in. high, when they develop offsets about the base, which may either be removed to form new plants, or allowed to remain and grow into a specimen like that in the Illustration. The flowers are large for the size of the plant, and they are developed freely in the apex of the stems in the early part of the summer. The tube is very short, naked, and completely hidden by the young mammae; sepals and petals in four series, the outer one pale purple, the inner of a deep purple colour; stamens very numerous, and the stigma has only four erect lobes. The plant was first described from examples cultivated in Berlin in 1843, but the flowers were not known till 1858. There are several varieties known, viz., *P. a. concolor*, which is distinguished by the whole of the flower being deep purple in colour; *P. a. pectinata* has larger scales (spine-tufts); and *P. a. cristata* is, as its name denotes a kind of cockscomb or crested form. They are all natives of Mexico.

CHAPTER XV.

THE GENUS OPUNTIA.

(The old Latin name used by Pliny, and said to have been derived from the city of Opus.)

HERE are about 150 species of Opuntia known, all of them natives of the American continent and the West Indies, though a considerable number have become naturalised in many other parts of the world. They are, with very few exceptions, easily distinguished from all other Cactuses by the peculiar character of their stems and spines; they are also well marked in the structure of their flowers. They vary in size from small, trailing, many-branched plants, never exceeding 6 in. in height, to large shrubs 8 ft. to 30 ft. high. (Humboldt states that he saw "Opuntias and other Cactuses 30 ft. to 40 ft. high.") Generally the branches are nearly flat when young, and shaped like a racquet or battledore; but in some species the branches are round (*i.e.*, in *O. cylindrica, O. subulata, O. arborescens*, &c.). All the kinds have fleshy stems, which ultimately become cylindrical and woody. At first they consist of fleshy joints, superposed upon one another, the joints varying considerably in size and shape. When young they bear small fleshy leaves along with the spine-tufts; but the former fall off at an early stage, whilst the spines are altered in length or number as the joints get old. In one or two kinds the spines fall away when the joints begin to harden, and in *O. subulata* the leaves are large and persistent.

The nature of the spines of Opuntias is of a kind that is not likely to be forgotten by anyone coming into contact with them. Every spine, from the tiny bristles, hardly perceptible to the naked eye, to the stout, needle-like spears which are found on the branches of some kinds, is barbed, and they are so very sharp and penetrating that even a gentle touch is sufficient to make them pierce the skin. Once in they are very difficult to get out; the very fine ones can only be shaved level with the skin, and left to grow out, whilst the larger must be cut out if they have penetrated to any depth. This horrid character in Opuntias, whilst rendering them disagreeable to the gardener, has been turned to good account in many of our colonies, where they are commonly used as fences. A good hedge of such kinds as *O. Tuna* or *O. horrida* is absolutely impassable to both man and beast, and as the stems are too watery to be easily destroyed by fire, their usefulness in this way could not be surpassed. As all the Opuntias will grow in the very poorest of soils, and even on bare rocks, and as they grow very rapidly, they have been largely employed in Africa, Australia, and India for fences. It is reported that when an island in the West Indies was divided between the French and English, the boundary was marked by three rows of *O. Tuna*.

The flowers of Opuntias are not, as a rule, particularly attractive. In many of the kinds they are large and well-formed, but the colours are tawny-yellow, greenish-white, or dull red. These plants cannot, therefore, be recommended for any floral beauty, although it is probable that the same flowers, on plants of less repulsive appearance than Opuntias are, as a rule, would be admired. There are a few exceptions to this in such species as *O. Rafinesquii, O. missouriensis*, and *O. basilaris*, which are compact and dwarf, and bear numerous large, brightly-coloured flowers. The fruits of Opuntias, or, at least, some of them, are edible, and to some palates they are very agreeable. We have tasted them, and consider they are mawkish and insipid—not much better than very poor gooseberries. Sir Joseph Hooker has compared them to Pumpkins. They are pear-shaped, with a thick, spine-covered rind, containing green, yellow, or red pulp, with small, hard seeds scattered through it.

The fruit of Opuntia differs in character and structure from the ordinary kind of fruit, such as apples, pears, &c. It consists of a branch, or joint, modified in form, and bearing on its flattened apex a flower, with the ovary buried in a slight depression in the fleshy joint. After becoming fertilised, the ovary grows down into the joint, and, ultimately the whole joint is changed into a succulent, juicy, often coloured "fruit." That this is the case has been proved by planting the unripe "fruit" of Opuntias in pots of sandy soil, and treating them as cuttings, when they have developed buds at the apex and roots at the base, ultimately forming plants.

The vitality in the branches of most of the species is very great, the smallest piece, as a rule, emitting roots and developing into a plant in a comparatively short time. The branches are soft, and easily broken, so that, in gathering the fruits, many pieces are broken off and cast aside; these soon grow into plants, and in a short time an extensive "colony" of Opuntias springs up where previously only one had been. The seeds, too, are a ready means of increase, being distributed by birds and other animals, which eat the fruits. In consequence of this free vegetative character, the Opuntias introduced into some of our colonies have become a pest almost as difficult to deal with as the rabbit scourge in Australia. In English gardens, however, there is no danger of Opuntias getting the upper hand. The adaptability of the majority of the kinds for cultivation under what may be termed adverse conditions for other plants, and the ease with which they may be propagated, render the management of a collection of these plants an easy matter. Amongst other Cactuses, Opuntias have a striking effect, and a selection of them should be grown in even the smallest collections. A few of them may be recommended specially as attractive plants for a sheltered, sunny rockery.

Cultivation.—The cultural requirements of the Opuntias may more conveniently be referred to under the description of each kind.

Propagation.—This entails no exceptional treatment; the numerous seeds contained in each fruit germinate freely if sown in sandy soil, and placed on a shelf in a warm house; and the smallest branches root quickly if planted in pots of open soil and kept in the Cactus-house. Large branches root just as freely as small ones. At Kew an enormous specimen, which had grown tall, and developed a thicket of branches too great for the house where it grew, was reduced most summarily by simply cutting off the head of branches and planting it in the ground where the original specimen had been. In a short time this "cutting" was well rooted, and made better growth than it had before the operation was performed.

As stocks for grafting, many of the more robust kinds of Opuntia are well adapted, and very singular-looking specimens may be obtained by making the most of this fact. One of the crested or monstrous forms, when grafted on a flat-stemmed kind, presents the queerest of appearances, looking like a large green cockscomb growing out of the top of a bladdery kind of stem. Equally odd combinations may be made by grafting a flat-stemmed kind on one whose stem is cylindrical. As all the kinds unite with the greatest ease, a taste for oddities among plants may easily be gratified by making use of Opuntias in this way. The time most favourable for the operation is spring—say, the month of April. For full information on how to graft Cactuses, see Chapter IV., on Propagation.

SPECIES.

O. arborescens (tree-like).—This species is known as the Walking-Stick or Elk-Horn Cactus, from its cylindrical, woody stems being made into very curious-looking walking-sticks (examples of which may be seen in the Museum at Kew), whilst the arrangement of the branches is suggestive of elk horns. Habit erect; joints cylindrical, branching freely, and forming trees from 8 ft. to 30 ft. high. Stems covered with oblong tubercles and tufts of long, needle-like spines, which give the plant a very ferocious aspect. Flowers on the ends of the young branches, 2 in. to 3 in. in diameter, bright purple in colour, developing in June. It is a native of Mexico, &c., and requires greenhouse or stove treatment. The skeletons of this species, as seen scattered over the desert places where it is wild, have a very singular and startling appearance. They stand in the form of trees, quite devoid of leaves, spines, or flesh, and, owing to the peculiar arrangement of the ligneous layers, nothing remains except a hollow cylinder, perforated with mesh-like holes, indicating the points where the tubercles and small branches had been. These skeletons are said to stand many years.

O. arbuscula (small tree). — Another of the cylindrical kinds, with a solid, woody trunk, about 4 in. through, and clothed with smooth, green bark; it grows to a height of 7 ft. or 8 ft. Branches very numerous, slender, copiously jointed, the ultimate joints about 3 in. long and = in. thick; they are slightly tuberculated, and bear tufts of spines nearly 1 in. long. Flowers 1= in. in diameter, produced in June; petals few, greenish-yellow, tinged with red. It is a native of Mexico, and requires stove treatment. A pretty plant, or, rather, a very remarkable one, even when not in flower, the thin branches, with their hundreds of long, whitish spines, being singular. Unfortunately, it is not easily grown.

O. arenaria (sand-loving). —Stems spreading, forming a tuft 3 ft. through and about 1 ft. high. Joints 1= in. to 3 in. long, and a little less in width, terete, with very prominent tubercles and numerous tawny bristles; upper spines 1 in. to 1= in. long, white, with a yellow point, shorter ones hair-like and curled. Flowers 2 in. in diameter, produced in May. Fruit 1 in. long, bearing a few short spines. Mexico. A strong-rooted plant, which should be grown in very loose, sandy soil. It would probably thrive best when planted out on a stage near the glass in a stove.

O. Auberi (Auber's).—An erect-growing plant, 8 ft. or more high, not unlike *O. Ficus-indica* in the form of its joints, but with long spines springing from the cushions, whereas the latter has none. The joints are oblong-ovate, glaucous-green, the cushions few and scattered; spines white, flattened, of various lengths. Flowers tawny yellow, small for the size of the plant. A native of Cuba, and requiring stove treatment. Being very brittle, this plant should be supported with stakes.

O. aurantiaca (orange).—A dwarf, cylindrical-stemmed kind, branching freely. Joints short, > in. in diameter; cushions of reddish spines, one about 1 in. long, the others shorter; bases of spines enveloped in white wool. Flowers bright orange, 2 in. to 3 in. across. This species is a native of Chili, whence it was introduced in 1824. It should be grown in a warm greenhouse all winter, and placed in a sunny position outside during summer.

O. basilaris (branching at the base); Fig. 76.—A dwarf, compact plant, of peculiar habit. Stem short, branching into a number of stout, obovate, often fan-shaped joints, which usually spring from a common base, and curve inwards, suggesting an open cabbage. Joints 5 in. to 8 in. long, about 1 in. thick, covered all over with dot-like cushions of very short, reddish spines, set in slight depressions or wrinkles. Flowers of a beautiful and rich purple colour, about 2= in. in diameter, and produced in May. This distinct plant is a native of Mexico, and is of recent introduction. Plants of it may be seen in the Kew collection. It is apparently easily kept in health in an ordinary stove temperature along with other Cactuses. It varies in the form of its joints and in its manner of branching, but it seems never to develop the joints one on the top of the other, as do most Opuntias. This species is certain to become a favourite when it becomes better known.

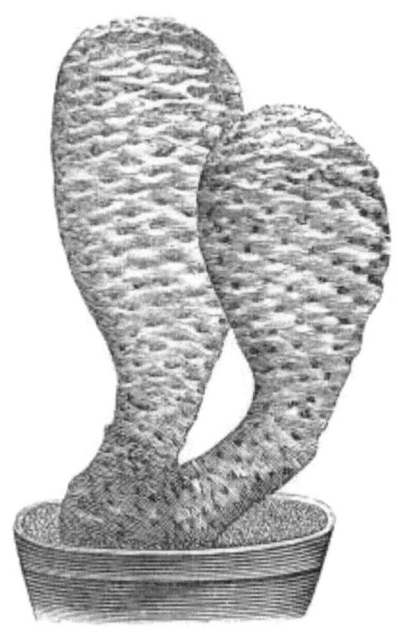

FIG. 76.—OPUNTIA BASILARIS

O. Bigelovii (Bigelow's). — A cylinder-stemmed, tall-growing plant, with a stout, woody stem, bearing a dense head of branches. Joints 2 in. to 6 in. long, 1 in. to 2 in. in diameter, light green, covered with small tubercles and little spine-cushions, with larger spines 1 in. long. When wild, the young joints are often shaken off by the wind, and cover the soil around, where they take root or stick to the clothes of the passers-by like burrs. Flowers not known. A native of Mexico, where it forms a tree 12 ft. high; it requires stove treatment. The skeleton of the trunk is a hollow cylinder, perforated with numerous holes, which occur in a regular spiral. The appearance of a full-grown specimen is very striking, the oval joints, thickly covered with long, needle-like spines, hanging in clusters, more suggestive of spiny fruit than branches.

O. boliviana (Bolivian); Fig. 77.—Stems 1 ft. high, erect, branching, and composed of roundish, pale green joints, with small, round tubercles, and long, white, flexible spines, sometimes as much as 4 in. in length; cushions about 1 in. apart. Flowers 1= in. across, yellowish. This is a fat, gouty-looking plant, from Bolivia, requiring stove treatment. It often assumes a yellow hue on the older joints, even when in good health.

FIG. 77.—OPUNTIA BOLIVIANA

O. brachyarthra (short-jointed); Fig. 78.—A dwarf-growing, singular-looking plant, with short, tumid joints from 1 in. to 2 in. long and wide, and nearly the same in thickness. The shortness of the joints, together with their growing on the top of each other, has been not inaptly compared to a jointed finger. Cushions very close together, composed of short, white and yellowish bristles, and stout, terete spines, 1 in. or more long, set on little tubercles. Flowers 1 in. in diameter, with about five sepals, eight or nine petals, and a five-rayed stigma; they are borne on the apices of the topmost joints. This species is worth growing on account of its peculiar stems and the length of its white spines. It is a native of New Mexico, and has been recently introduced to Kew, where it is cultivated among the hardy kinds, and also in the greenhouse.

FIG. 78.—OPUNTIA BRACHYARTHRA

O. braziliensis (Brazilian).—The peculiar habit and mode of growth at once distinguish this species. It rises with a perfectly straight, erect, slender, but firm and stiff, round stem, to a height of from 10 ft. to 30 ft., tapering from the base upwards, and furnished all the way up with short, horizontal branches, spreading about 3 ft. all round, like an immense candelabrum. Spines long, subulate, very sharp, ash-coloured, in clusters. Joints broadly oblong, margins wavy; they resemble leaves, or the thin, leaf-like joints of a Phyllocactus, with the addition of long, whitish spines on both sides. Flowers 1= in. in diameter, lemon-yellow, very freely produced on the younger joints during May and June. Fruit as large as a walnut, spiny, yellow when ripe. This species is a native of Brazil, whence it was introduced in 1816. It may be recommended for large, airy houses, as it grows freely, and forms a striking object when arranged with foliage and flowering plants of the ordinary kind. Its fruits are edible.

O. candelabriformis (candelabrum-shaped).—Stems erect, 5 ft. to 8 ft. high; joints flat, almost circular, about 6 in. in diameter, glaucous-green, densely clothed with numerous cushions of white, bristle-like spines, a few in each cushion being long and thread-like. Flowers not known on cultivated plants. This sturdy species is a native of Mexico, and succeeds well if planted on a little rockery or raised mound in a warm house, where, properly treated, it branches freely, and forms a dense mass of circular joints. It is one of the most useful of the larger Opuntias for cultivation in large houses.

O. clavata (club-shaped). — Stem short; joints club-shaped, 2 in. long and 1 in. wide, narrowed almost to a point at both ends. Cushions < in. apart, composed of numerous spines, varying from short and bristle-like to 1 in. in length, stout, flattened, and spear-like. Leaves < in. long. Flowers yellow, 1= in. across. Fruit 1= in. long, lemon-yellow when ripe, and covered with stellate clusters of white, bristle-like spines. New Mexico, 1854. A stove species, remarkable for the strength and form of its central spines, which are spear or dagger-shaped.

O. cochinellifera (cochineal-bearing); Bot. Mag. 2742. — An erect-growing plant, attaining a height of 9 ft. or more, and branching freely, the older parts of the stem and branches being woody and cylindrical; young joints flat, oblong-ovate, varying in length from 4 in. to 1 ft., deep green, rather soft and watery, spineless, the cushions distant, and sometimes bearing a few very short bristles. Flowers at the extremities of the branches, 1= in. long, composed of numerous imbricating, scale-like petals, curving inwards, and coloured crimson. Fruit flat-topped, 2 in. long, red; pulp reddish; seeds black. It is a native of tropical South America, whence it was introduced in 1688. It requires stove treatment, and blossoms in August. This is one of the most useful of the genus, on account of its being the kind chiefly employed in the cultivation of cochineal. It is one of the easiest to manage, requiring only a rather dry atmosphere, plenty of light, and a temperature not lower than 50 degs. in winter. Syn. *Nopalea cochinellifera*.

O. corrugata (wrinkled). — Stem not more than 2 ft. high; joints cylindrical, wrinkled all over, about 2 in. long, covered with cushions of white hair or bristle-like spines. Flowers 1= in. across, reddish-yellow, produced in August. A native of Chili, whence it was introduced in 1824. It may be grown in an ordinary greenhouse, on a shelf near the glass, and exposed to full sunshine.

O. curassavica (Curassoa); Pin-pillow. — Branches spreading; joints cylindrical or club-shaped, dark green, bearing numerous cushions of woolly bristles, and long, white, very sharp-pointed spines. Flowers 3 in. across, greenish-yellow, borne on the young joints in June. Introduced from Curassoa in 1690. A free-growing plant under favourable conditions, and one requiring stove treatment. It has been cultivated in gardens almost as long as any species of Cactus. There are several varieties of it known, differing from the type in habit, length of spine, or shade of colour in the flower.

O. cylindrica (cylindrical).—Stem and joints cylindrical, the latter covered with spindle-shaped tubercles, each one crowned with a tuft of fine, hair-like, whitish spines, one or two in each tuft being stiff, and sharp as needles. The leaves are fleshy, cylindrical, 1 in. or more long, and they remain on the joints longer than is usual in Opuntias. Flowers crowded on the ends of the branches, each 1 in. in diameter, scarlet; they are developed in June. This plant is said to grow to a height of 6 ft. or more in its native habitat, but under cultivation it is rarely seen more than 3 ft. high; it was introduced in 1799. It is handsome and distinct enough to be worth growing. It requires stove or greenhouse treatment, but rarely flowers under cultivation.

O. c. cristata (crested).—A dwarf, cockscomb-like variety, with the leaves and white hairs growing all along the wrinkled top of the comb. It is a very singular example of a "monster" Cactus. It requires stove treatment.

O. Davisii (Davis'); Bot. Mag. 6652.—Stems somewhat horizontal, not exceeding 1= ft. in height; joints 4 in. to 6 in. in length, and about = in. in thickness; wood dense, and hard when old; tubercles not prominent, bearing cushions of very slender bristles, forming a kind of brush, from amongst which the spines spring. The longest spines are 1= in., and they are covered with a loose, glistening sheath. Flowers 2 in. in diameter, greenish-brown. The plant is a native of New Mexico, and was introduced in 1883. It forms a compact, shrubby little plant if grown in an intermediate house during winter, and placed in the open in full sunshine during summer. It was flowered for the first time in England in 1883, and although not what we should call an attractive plant, in America it is described as being "a well-marked and pretty species." It is named after Jefferson Davis, the American statesman.

O. decumana (great-oblong). This is the largest-growing species in cultivation. At Kew it is represented by a plant 12 ft. high (it would grow still taller if the house were higher). It has a hard, woody, brown-barked stem, bearing an enormous head of very large, elliptical, flat joints, 12 in. to 20 in. long, and about 1 ft. broad, smooth, grey-green, with a few scattered cushions of very tiny bristles, and sometimes, though rarely, a spine or two. Flowers large, orange-coloured, produced in summer. Fruit oval, 4 in. long, spiny, brownish-red, very watery when ripe; flesh red, sweet. A native of Brazil, and requiring stove treatment. This is said to be what is known in Malta as the Indian Fig. The plant is chiefly interesting here on account of the extraordinary size of the joints.

O. diademata (diademed).—A small, remarkable, and extremely rare little species, with a short, erect stem, composed of globose, superposed joints, grey-green in colour, and very succulent. The topmost joint is pear-shaped, with a tuft of whitish hair and spines on the apex, out of which the new growth pushes. Cushions large, about 1 in. apart, furnished with a tuft of short, grey hairs and short spines, with a large one at the base. The character of this large spine is exceptional, being broad, flat, cartilaginous, whitish, and curving downwards. On healthy large examples these spines are 2 in. long, and nearly < in. wide at the base. Flowers and fruit not known. Native of Mendoza (La Plata). This little plant requires to be cultivated in a warm greenhouse or stove, but it grows very slowly. It is certainly a most interesting Cactus; examples of it may be seen at Kew, where there is a plant which, although over ten years old, is only 4 in. high. Syns. *O. platyacantha* and *Cereus syringacanthus*.

O. Dillenii (Dillenius'); Fig. 79.—An erect-growing, robust species, attaining a height of 15 ft., with flattened, ovate joints, about 5 in. long by 3 in. broad. Cushions composed of short, white, hair-like bristles, and numerous long, stout, yellow spines. Flowers yellow, tinged with red, 4 in. in diameter, freely produced on the ends of the youngest joints all summer. Fruits similar to those of *O. Ficus-indica*. A native of the West Indies, now naturalised in all warmer parts of the world. In India it is so plentiful and widespread that Roxburgh, an Indian botanist, said it was a native. In India, its fruits are eaten by the poor natives, and it is often planted as a hedge. It is also a great pest in the open lands of that country, and large sums are annually expended in cutting it down and burying it. This species, which requires warm greenhouse treatment, is also employed in the cultivation of cochineal.

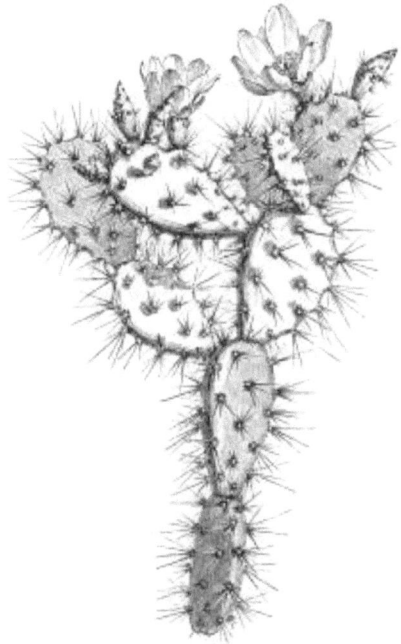

FIG. 79.—OPUNTIA DILLENII

O. echinocarpa (spiny-fruited).—A low, straggling shrub, not exceeding 1= ft. in height. Joints cylindrical, from 1 in. to 3 in. long, less than 1 in. thick. Cushions of rather coarse bristles and numerous spines, from = in. to 1 in. in length. Flowers 2 in. in diameter, yellow, produced in summer. Fruit short, depressed, almost saucer-shaped, and bearing spines nearly 1 in. long. A native of Colorado, &c. It requires stove treatment. The variety *major* has stems 4 ft. high, joints 8 in. to 10 in. long, and long, sheathed spines. This species is closely related to *O. Bigelovii* and *O. Davisii*.

O. Emoryi (Emory's).—A prostrate, spreading plant, less than 1= ft. high. Joints cylindrical, curved, 4 in. long, 1= in. thick. Tubercles very prominent, longitudinally attached to the stem, the apices crowned with pea-shaped cushions of short bristles, and numerous radiating spines, some of which are fully 2 in. long, very strong and needle-like. Flowers 2= in. in diameter, sulphur-yellow, tinged with purple, produced in August and September. Fruit 2= in. long and 1 in. thick, covered with cushions of bristles and spines. A native of Mexico, on dry, sandy soils, where its prostrate stems, clothed with powerful spines, form a hiding-place for the small animals, snakes, &c. Stove or warm greenhouse treatment is best for this species.

O. Engelmanni (Engelmann's).—A stout, coarse-looking plant, 6 ft. high, with woody stems and large, flat, green joints, 1 ft. long and 9 in. in diameter. Cushions 1= in. apart, composed of coarse bristles, and one or two spines over 1 in. long, and pointing downwards. Flowers 3 in. in diameter, yellow, produced in May and June. Fruit nearly round, 2 in. long, purplish both in rind and pulp, the latter rather nauseous to the taste. Mexico. This is a greenhouse plant which grows freely and flowers annually under cultivation. It is very similar to *O. monacantha*, a much better known species. According to American botanists, it is probably the most widely spread of the whole Cactus tribe.

O. Ficus-indica (Indian Fig); Fig. 80.—Branches erect, 8 ft. to 12 ft. high; joints flat, oval or obovate, about 1 ft. long by 3 in. in width, and 1 in. in thickness. Stems hard and woody with age. Cushions 1= in. apart, composed of short, yellowish bristles, and very rarely one spine. Flowers 3 in. to 4 in. across, sulphur-yellow, produced all through the summer. Fruit 3 in. to 4 in. long, pear-shaped, covered with tufts of bristles, white, yellow, or red when ripe. It is a native of Central America, whence it was introduced about 300 years ago. It is now widely spread, in tropical and temperate regions all over the world. In many parts it is cultivated for the sake of its fruits, which in some of our colonies are used for dessert. In England it must be protected from damp and cold; it is, therefore, best cultivated in a sunny greenhouse during winter, and placed outside in a position exposed to full sunshine all summer. Tenore, an Italian botanist, named this species *O. vulgaris*, and this mistake has led others to consider the North American *O. vulgaris* (true) and *O. Ficus-indica* as one and the same species.

FIG. 80.—FRUITING BRANCH OF OPUNTIA FICUS-INDICA

O. filipendula (hanging filaments); Fig. 81.—Stems prostrate, about 1 ft. high, spreading; joints flat, round or oval, about 3 in. long, often less, milky-green in colour. Cushions = in. apart, composed of a little tuft of white woolly hair, a cluster of erect, rather long bristles, like a small shaving-brush, and all pointing upwards; spines usually only one in each cushion, and this is slender, deflexed, white, and from 1 in. to 2 in. long. Sometimes the joints are wholly spineless. Flowers 2= in. in diameter, purplish, very handsome, produced in May and June. Fruit not known. The roots of this species bear tubers often 1 in. in thickness, and several inches in length, and these tubers will grow into plants if severed and planted. It requires stove treatment. Native country, Mexico.

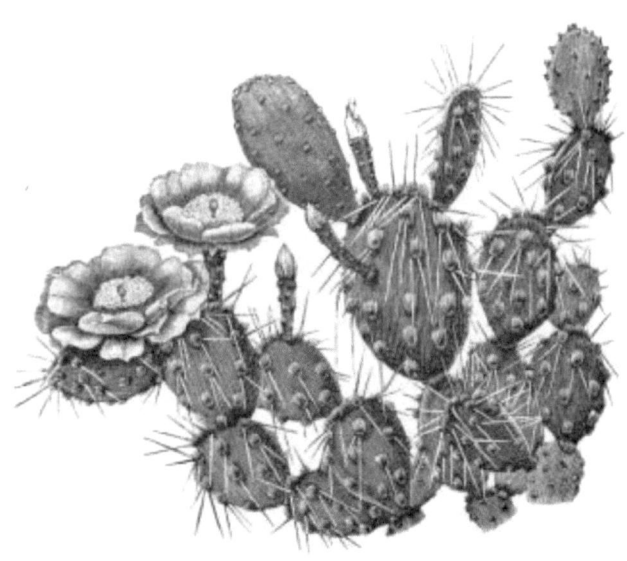

FIG. 81.—OPUNTIA FILIPENDULA

O. frutescens (shrubby).—A thin-stemmed, copiously-branched species. Joints almost continuous, like ordinary branches, from 2 in. to 6 in. long, the thickest not exceeding < in. Cushions on raised points or tubercles, each consisting of a small tuft of hair, inclosed in a row of bristles, and one long, central spine, often exceeding 2 in. in length. When young, the spines are inclosed in a thin, bony sheath. Flowers scattered along the younger branches, 1 in. across, greenish-yellow, borne in June. Fruit 1 in. long, pear-shaped, smooth, scarlet, with tufts of bristles all over it, and a depression in the apex. Mexico. This forms an interesting pot-plant when properly cultivated. It should be grown in a warm greenhouse.

O. Grahami (Graham's).—This is one of several species of Opuntia which are remarkable in having thick, fleshy roots, not unlike those of the Dahlia. The joints are 2 in. long and 1 in. in diameter, cylindrical, with adpressed tubercles, = in. or more long, each tubercle bearing a tuft of long, straight, radiating spines. Flowers 2 in. across, yellow, borne on the ends of the ripened joints in June. Fruits 1= in. long and > in. wide, covered with stellate clusters of short, bristle-like spines. This plant is a native of Mexico, and is a recent introduction. From the nature of its roots, which are no doubt intended to serve as reservoirs for times of extreme drought, it should be grown in well-drained, sandy soil, and kept quite dry all winter. It requires stove treatment.

O. horrida (horrid).—An erect, stout-stemmed plant, with flattened, green joints, about 5 in. long by 3 in. wide. Cushions 1 in. apart, composed of short, reddish bristles, and long, tawny red spines, about eight in each cushion, and of a peculiarly ferocious appearance—hence the specific name. The stoutest spines are 3 in. long, and are sharp and strong as needles. This species (which is probably a native of Mexico) is deserving of a place in collections of Cactuses because of the character of its spines. Probably it is only a variety of *O. Tuna*. It requires warm-house treatment.

O. hystricina (porcupine-like).—This beautiful species was discovered in the San Francisco Mountains mixed with *O. missouriensis*, to which it is nearly allied. It is spreading in habit, the joints 3 in. to 4 in. long and broad; cushions = in. apart, rather large, with numerous spines, varying in length from = in. to 4 in., and short, yellowish bristles. Flowers large, yellow. Fruit 1 in., long, spiny. This plant is not known in English collections, but it is described by American botanists as being attractive and a free grower. As it is found along with *O. missouriensis*, it ought to prove hardy in England.

O. leptocaulis (slender-stemmed).—This little Mexican species is chiefly remarkable for its fragile, numerous, twig-like joints, thickly dotted with tubercles and numerous spirally-arranged cushions of reddish bristles, with long, grey spines. It does not flower under cultivation. Requires stove treatment.

O. leucotricha (white-haired).—An erect-stemmed kind, with flattened joints, ovate or oblong in shape, and bearing numerous cushions, = in. apart, of short bristles, with a large, central spine, and a few others rather shorter. When young these spines are rigid and needle-like; but as they get older they increase in length, and become soft, and curled like stiff, white hair. Young plants are noticeable for their small, subulate leaves of a bright red colour, whilst old examples are almost as interesting as the Old Man Cactus (*Pilocereus senilis*), the long, white, hair-like spines of the Opuntia hanging from the older joints in much the same manner as they do from the upper part of the stem of the Pilocereus. Flowers yellow, produced in June. This species is a native of Mexico, and requires stove treatment. Seeds of this, and, indeed, of a large proportion of the cultivated Opuntias, may be procured from seedsmen, and as they germinate quickly, and soon produce handsome little plants, a collection of Opuntias is thus very easily obtained.

O. macrocentra (large-spurred).—A flat-jointed species, growing to a height of 3 ft.; the joints large, almost circular, thinly compressed, and usually purplish in colour. Cushions about 1 in. apart, with spines often 3 in. long, of a greyish colour, and generally pointing downwards. Flowers 3 in. across, bright yellow; they are developed in May and June, on the upper edges of the youngest joints. This plant is a native of Mexico; it is at present rare, but the unusual colour of the joints, its compact, freely-branched habit, the extraordinary length of its spines, and the size of its flowers, ought to win for it many admirers. It is easily grown if kept in an intermediate house. Plants of it may be seen in the Kew collection.

O. macrorhiza (large-rooted); Figs. 82, 83.—In this Texan species we have a combination of the principal characters for which the genus Opuntia is remarkable: The thick, fleshy roots, which are a supposed source of food, and which look like potatoes; the cylinder-shaped older stems, and the flattened, battledore-like joints; the tufts of bristles on the stems, and deciduous, longer spines on the joints; the large, beautiful, yellow flowers; and the small leaves on the newly-formed joints. In habit and flowers this kind resembles *O. Rafinesquii*; and if not quite hardy in England, it is nevertheless sufficiently so to thrive in any sunny position where it would be protected from frost and excessive wet. The accompanying illustrations represent the characters of this species so well that further description is not needed. The flowers are developed in early summer.

FIG. 82.—FLOWERING BRANCHES OF OPUNTIA MACRO-RHIZA

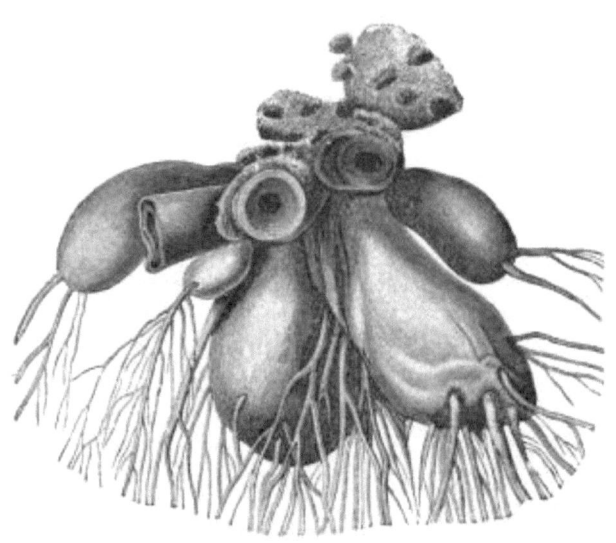

FIG. 83. ROOTS OF OPUNTIA MACRORHIZA

O. microdasys (small, thick).—This is a handsome little Mexican plant. Its flattened joints, which are nearly circular in outline, are thickly covered with little cushions of bright orange-yellow bristles, the cushions being so close together that the short bristles almost hide the green joints from view. The stems are semi-decumbent, and they branch somewhat freely. Flowers not seen. It thrives in a warm greenhouse temperature. The best examples of this pretty Opuntia are grafted on a robust-growing kind, the stock being about 1 ft. long, and the scion forming a compact head of pretty, healthy-looking joints. Treated in this way, this species is most interesting and attractive. It may also be grown on its own roots. There is a variety of it, named *rufida*, in which the bristles are reddish-brown.

O. missouriensis (Missouri).—A stout, prostrate kind, forming large, spreading masses under favourable conditions. Joints broad, flattened, obovate, about 4 in. long by 2 in. wide, light green; spine-cushions less than 1 in. apart, and composed of numerous small, white spines, with from one to four longer ones; these latter fall away when the joints get old. Leaves very short, with a little wool about their bases. Flowers 3 in. in diameter, appearing from May onwards; petals yellow, dashed with rose, sometimes wholly rose-coloured or brick-red. Stamens deep red; pistil yellow, with a conical stigma. Fruit nearly round, spiny, about 2 in. long. A native of Wisconsin, and westward to the San Francisco Mountains; introduced in 1814. This species is as hardy as *O. Rafinesquii*, and thrives under similar treatment. It has stood 22 degs. of frost without suffering, requiring only protection from rain in winter. In North America it forms large, spreading masses on gravelly hillsides, and is much dreaded by travellers, and especially by horses; there it is usually covered with snow from Christmas to the following May.

O. monacantha (one-spined).—A tall, robust plant, not unlike *O. Dillenii* in general habit. It has flat, large joints, oblong or ovate in outline, rather thinly compressed, and bearing grey cushions over 1 in. apart, with a solitary spine, 1= in. long, springing from the centre of each cushion, and pointing downwards. Flowers sulphur-yellow, 2= in. across, borne on the last-ripened joints in May, and abundant on well-grown plants. Fruits ovate, 2 in. long, green, with tufts of short, brown bristles; pulp edible. The species is a native of Brazil, but is now common in many tropical and sub-tropical countries. It is a free-growing kind, soon forming a large specimen if planted in a bed of old brick-rubble, or other light, well-drained soil, and kept in warm greenhouse temperature.

O. nigricans (blackish); Bot. Mag. 1557.—Stem stout, erect, becoming hard and woody when old. Joints flat, oval in outline, 5 in. to 8 in. long. Cushions 1= in. apart, composed of short reddish-brown bristles and two or three long stout spines, which are yellow when young, but almost black when ripe. Flowers produced on the young, ripened joints, orange-red, about 3 in. across and developed in August and September. Fruit pear-shaped, rich crimson when ripe. Introduced from Brazil in 1795. This well-marked species thrives in a warm greenhouse. It branches freely, and has a healthy aspect at all times. It is represented at Kew by very large specimens; one of them, which was recently cut down, had a stem 12 ft. high and an enormous head of dark, green joints. Its head was planted as a cutting.

O. occidentalis (Western). —Stem stout, woody, with innumerable branches, wide-spreading, often bent to the ground. Joints 9 in. to 12 in. long by about 6 in. broad, flattened, as many as 100 on one plant. Cushions nearly 2 in. apart, with small, closely-set bristles and straight spines from = in. to l= in. long. Flowers produced in June on the ripened joints, nearly 4 in. in diameter, orange-yellow. Fruit 2 in. long, "very juicy, but of a sour and disagreeable taste." This is an exceptionally fine plant when allowed sufficient space to develop its enormous branches and joints; it is a native of the Western slopes of the Californian mountains. It should be planted in a bed of rough, stony soil, in a dry greenhouse. Possibly it is hardy, but it does not appear to have been grown out of doors in England.

O. Parmentieri (Parmentier's).—Stem erect. Joints cylindrical, "like little cucumbers." Cushions about 1 in. apart, arranged in spiral rows, and composed of short, reddish bristles, with two or three straw-coloured spines, 1 in. long. Flowers reddish, small. The plant is a native of Paraguay, and is rarely heard of in cultivation. It requires stove treatment.

O. Parryi (Parry's).—Stem short. Joints club-shaped, 4 in. to 6 in. long, very spiny, the cushions elevated on ridge-like tubercles. Bristles few, coarse, and long. Spines very numerous, varying in length from < in. to 1= in.; central one in each cushion much the broadest, and flattened like a knife-blade, the others being more or less triangular. Flowers yellowish-green, on the terminal joints, which are clothed with star-shaped clusters of bristle-like spines, the flowers springing from the apex of the joint, and measuring 1= in. across. A native of Mexico, where it grows on gravelly plains. This distinct plant is in cultivation at Kew, in a warm greenhouse, but it has not yet flowered.

O. Rafinesquii (Rafinesque's); Fig. 84.—A low, prostrate, spreading plant, seldom exceeding 1 ft. in height, the main branches keeping along the ground, the younger ones being erect. The latter are composed of flat, obovate joints, 4 in. to 5 in. long by 3 in. in width, fresh green in colour; spines very few, mostly only on the upper edge of the last-made joints, single, or sometimes two or three from each spine-cushion, 1 in. long, straight, whitish, soon falling off; cushion composed of very fine reddish bristles and whitish wool; leaves very small, falling early. The branches become cylindrical and woody with age. Flowers 2 in. to 4 in. in diameter, bright sulphur-yellow, with a reddish tint in the centre; in form they are like a shallow cup, the numerous stamens occupying the middle. They are produced in great abundance on the margins of the youngest joints, as many as fifty open flowers having been counted on a single specimen at one time. Fruit pear-shaped, 1= in. to 2 in. long, naked, edible, somewhat acid and sweetish. The flowering season is from July to September; the native country, Wisconsin to Kentucky, and westward to Arkansas and Missouri. This species, introduced about twenty years ago, has only recently been brought prominently before English gardeners. It is a very ornamental and interesting plant for outdoor cultivation, and when once established gives no trouble. For the first year or two after planting it requires watching, as, until the basal joints harden and become woody, they are liable to rot in wet weather. A large-flowered form, known as *grandiflora*, is cultivated in American gardens.

FIG. 84.—OPUNTIA RAFINESQUII

O. rosea (rose-coloured); Fig. 85.—Stem erect, branching freely. Joints varying in length from 2 in. to 6 in., not flattened, with ridge-like tubercles, bearing on their points small cushions of very fine bristles and tufts of pale yellowish spines about = in. long, and all pointing upwards. Flowers on the ends of the ripened growths of the year, usually clustered, 2 in. across, bright rose-coloured; they are developed in June. A rare species from Brazil, and one which, as the illustration shows, is both distinct and handsome enough to be classed amongst the most select. It requires a stove temperature.

FIG. 85.—OPUNTIA ROSEA

O. Salmiana (Prince Salm-Dyck's).—Stem erect, branching freely, the branches at right angles to the stem. Joints from 1 in. to 6 in. long, cylindrical, smooth, = in. in diameter, clothed with small cushions of soft, short bristles, and one or two longish spines. Flowers produced in September, 2 in. across, yellow, streaked with red, of short duration. Fruit egg-shaped, 1 in. long, crimson. This species is a native of Brazil, whence it was introduced in 1850. It requires to be grown in an intermediate house. It is a charming little Cactus, and quite exceptional among Opuntias in the colour and abundance of its flowers, and in the rich colour of its numerous fruits, which usually remain on the plant several months. The plant, too, has the merit of keeping dwarf and compact. The small joints separate very easily from the branches, and every one of them will root and grow into a plant. There is something very remarkable in the development of the fruits of this kind. A small branch, or joint, grows to its full length, and a flower-bud appears in the apex. If examined at this stage, it will be seen that the ovary occupies only a very shallow cavity in the top of the branch. After flowering, this ovary grows into the branch, and ultimately the whole branch is transformed into a pulpy fruit, with the seeds scattered all through the pulp. This peculiarity is well shown in *O. salmiana*, and the development of the fruit can be very easily watched. Many of the small branches do not flower, although they change to a red colour like the fruits.

O. spinosissima (very spiny).—Stem erect, woody. Joints very flat and thin, deep green, ovate or rotund, from 6 in. to 1 ft. long. Cushions 1 in. apart. Bristles very short. Spines in clusters of about five, the longest 2 in. in length, brownish-yellow. Flowers reddish-orange, small, usually only 2 in. across, produced in June. A native of South America; naturalised in many parts of the Old World. The stem becomes cylindrical with age, and sometimes is devoid of branches for about 5 ft. from the ground. The plant requires stove treatment. Probably this kind is only a form of *O. Tuna*.

O. subulata (awl-shaped). —Stem erect, cylindrical, even below, channelled and tubercled above, about 2 in. in diameter. Joints long and branch-like, with tufts of short, white hair on the apices of the tubercles, and one or two white, needle-like spines from = in. to 1 in. long. At the base of each tuft, from the apex to 1 ft. or more down the younger branches, there is a fleshy, green, awl-shaped leaf, from 2 in. to 5 in. long. Ultimately the leaves and spines fall away, the tubercles are levelled down, and the mature stem is regular and cylindrical, with tufts of white setae scattered over it. Flowers small, produced in spring; sepals 2 in. long, green, deciduous; petals small, dull purple, usually about eight in each flower. Fruit pear-shaped, 4 in. long; seeds very large, nearly = in. long and wide. This handsome South American species was the subject of an interesting communication to the *Gardeners' Chronicle*, in 1884, from Dr. Engelmann. It had previously been known as a Pereskia from the fact of its leaves being persistent and very large. In its leaves, flowers, and seeds, *O. subulata* is one of the most interesting of the genus. It is easily grown in a warm greenhouse, and deserves a place in all collections of Cactuses.

O. Tuna (native name); Fig. 86.—An erect-stemmed, flat-jointed, robust-growing species. Joints ovate, 4 in. to 9 in. long, with cushions 1 in. apart, composed of short, fulvous bristles, and several long, needle-shaped, unequal, yellowish spines. Flowers borne on the upper edges of the last-ripened joints, 3 in. across, reddish-orange, produced in July. Fruit rich carmine, about 3 in. long, pear-shaped. The plant is a native of the West Indies, &c., and was introduced in 1731. It has already been stated, under *O. spinosissima*, that there is a close similarity between that species and *O. Tuna*. We suspect, also, that *O. nigricans* is another near relation of these two. They are much alike in all characters, and they require the same treatment. *O. Tuna* has been seen as much as 20 ft. in height.

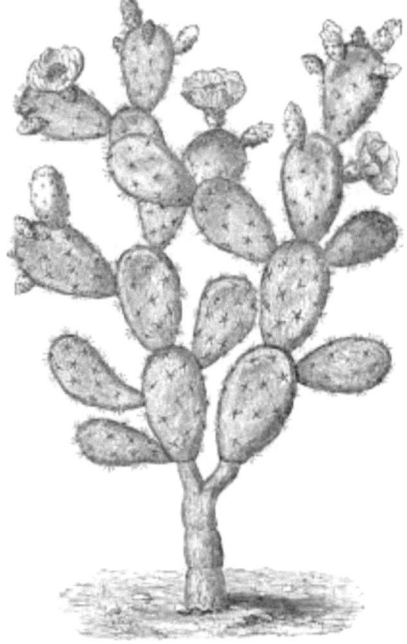

FIG. 86.—OPUNTIA TUNA

O. tunicata (coated-spined).—Stem sub-erect, cylindrical. Joints club-shaped, variable in length, about 2 in. in diameter. When young the surface is broken up into numerous oblong tubercles, each bearing a small cushion of whitish, short hairs, and about half a dozen white spines, unequal in length, the longest stout, and inclosed in a hard sheath, which becomes broken and ragged when old. Flowers not known. A native of Mexico, and introduced in 1840. It requires stove treatment.

O. vulgaris (common); Bot. Mag. 2393.—A low, prostrate, spreading plant. Joints short, oval, flattened, thicker than in *O. missouriensis*, 3 in. long by about 2 in. broad. Spine-cushions > in. apart; tufts very small, with, occasionally, a long spine. Leaves fleshy, very small. Flowers 2 in. across, pale sulphur-yellow. Fruits nearly smooth, 1= in. long, brown when ripe, with a strong disagreeable odour. The flowers are produced freely in June. The plant grows wild in Mexico, and extends up to New York, usually near the coast. It is now common in many parts of Europe, where it has become naturalised. In Madeira it has taken possession of all waste land, and is perfectly at home there. In England it was cultivated by Gerard nearly 300 years ago. It grows rapidly if planted in stony soil, in a position exposed to full sunshine, where it will creep along the ground, and root all along its stems, which rarely get elevated more than 6 in. from the ground. This species and *O. Ficus-indica* are confused by some authors, owing, no doubt, to the name *O. vulgaris* having been given by a botanist to the latter, which is a much larger and very different-looking plant. *O. vulgaris* is capable of withstanding our winters out of doors.

O. Whipplei (Captain Whipple's).—Stem usually prostrate, with slender, elongated branches, which are cylindrical when old, broken up into short joints when young. Joints varying in length from 2 in. to 1 ft., less than 1 in. in diameter. Cushions small, round. Spines white, variable in number, and arranged in tufts on the ends of the tubercles, one being 1 in. long, the others shorter. Flowers nearly 2 in. in diameter, red, borne in a cluster on the ends of the last-ripened joints in June. Fruit 1 in. long, with a cavity in the top. A compact, Mexican species, with crowded branches, and very free-flowering. It requires stove treatment. *O. Whipplei* is related to *O. arborescens*, from which, however, it is easily distinguished by the latter having a stout central spine and numerous radiating ones.

Of the 150 species of Opuntia known, about one-third have been selected for description here, and amongst these will be found all the best-marked kinds in the genus, and most of those of which we have any knowledge. Botanists find good specific characters in the size and structure of the seeds, in the character of the fruits, &c.; but for horticultural purposes these are of little or no value.

CHAPTER XVI.

THE GENUS PERESKIA.

(Named in honour of Nicholas F. Peresk, a botanist of Provence.)

HE thirteen species included in the genus Pereskia differ so markedly from all other kinds of Cactus, that at first sight one can scarcely believe they are true Cactuses, closely related to Cereus and Epiphyllum. They have erect or trailing stems and branches, and usually form dense, large bushes; the branches are woody and thin, and bear large, laurel-like leaves, which remain on the plants several years—so that they may be termed evergreen. They have, however, the spine-cushions, the tufts of woolly hair and stout spines, and the floral characters which distinguish Cactuses from other plants; they are also succulent, the leaves and young branches being soft and fleshy. They appear to have the same peculiar provision for enabling them to bear long periods of drought without suffering that characterises the more familiar forms of Cactuses. The development of the spines in this genus is different from what takes place in all other spiny plants of this order. In the latter the spines are stoutest and most numerous on the younger parts of the plant, the older or woody parts being either spineless, through having cast them, or much less spiny than when they were younger. Thus, in Opuntia we find few or no spines on the old parts of the stems of even such species as *O. horrida*, *O. nigricans*, &c. In Echinocactus, too, the spines about the base of old plants are much fewer, if not entirely cast off, than on the upper part. In Pereskia the contrary is

the case. Taking *P. aculeata* as an example, this is best known in gardens as having branches about as thick as a goose-quill, with ovate leaves, at the base of which there is a pair of curved spines, < in. long, and shaped like cats' claws. But this plant when it gets old has a stem 3 in. in diameter, and clothed down to the ground with cushions of spines fixed firmly in the bark, each cushion composed of from twenty to fifty spines, and each spine 1 in. or more in length. From two to six new spines are developed in the centre of each healthy cushion annually. It would be absolutely impossible for any animal to climb an old stem of a Pereskia. In *P. Bleo* the spines are 2 in. long, and the cushions are much larger.

The flowers of Pereskias are borne singly or in panicles, at the ends of the young, ripened branches. In shape, each flower may be compared to a single Rose, the petals being flat and spreading, and the numerous stamens forming a compact cluster in the centre. The stigma is erect, and divided at the top into four or more rays. The fruit is a berry shaped like a Gooseberry, and covered with minute clusters of short bristles.

All the species are found in tropical America and the West Indies.

Cultivation.—Although several of the kinds of Pereskia are sufficiently ornamental to be deserving of a place in gardens as flowering plants, yet they are rarely cultivated—in England, at least—for any other purpose than that of forming stocks upon which Epiphyllums and other Cacti are grafted. Only two species are used, viz., *P. aculeata* and *P. Bleo*, the former being much the more popular of the two; whilst *P. Bleo*, on account of the stoutness of its stems, is employed for only the most robust kinds of grafts.

Propagation.—Both the above-named species may be propagated to any extent, as every bit of branch with a leaf and eye attached is capable of rooting and soon forming a stock. The practice among those who use Pereskias as stocks for Epiphyllums is as follows: Cuttings of *P. aculeata* are planted in sandy soil, in boxes, and placed on a shelf in a stove till rooted. In about a month they are ready to be planted singly in 3 in. pots, any light soil being used; and each plant is fastened to a stake 1 ft. long. They are kept in a warm, moist house, all lateral shoots being cut away, and the leader encouraged to grow as tall as possible in the year. From December the plants are kept dry to induce the wood to ripen, preparatory to their being used for grafting in February. Stocks 9 in. or 1 ft. high are thus formed. If taller stocks are required, the plants must be grown on till of the required length and firmness. Large plants may be trained against a wall or along the rafters in a warm house; and when of the required size, the branches may be spurred back, and Epiphyllums, slender Cereuses, and similar plants, grafted upon them. In this way very fine masses of the latter may be obtained in much less time than if they were grown from small plants.

SPECIES.

P. aculeata (prickly); West Indian or Barbados Gooseberry.—Stem woody, more or less erect, branching freely, and forming a dense bush about 6 ft. high. Young branches leafy; old ones brown, leafless, clothed with large cushions of long, stout, brown spines, sometimes 2 in. in length. Leaves alternate, with very short petioles, at the base of which is a pair of short spines, and a small tuft of wool in the axil; blade 3 in. long by 2 in. broad, soft, fleshy, shining green. Flowers semi-transparent, white, in terminal panicles; sepals and petals > in. long by < in. wide; stamens in a large, spreading cluster, white, with yellow anthers. Ovary covered with small cushions of short bristles, with sometimes a solitary spine in the centre of each cushion. Fruit 1 in. long, egg-shaped, red, edible. There is a large plant of this in the Succulent House at Kew which flowers almost annually, but it has never ripened fruits. In the West Indies it is a very common shrub, whilst at the Cape of Good Hope it is used for fences—and a capital one it makes.

P. a. rubescens (reddish).—This variety has narrower, longer leaves, which are glaucous-green above and tinged with red below; the spines on the old stems are shorter and more numerous in each cushion. This requires the same treatment as the type.

P. Bleo (native name); Fig. 87.—A stout, branching shrub, having an erect stem, 3 in. or more in diameter, with green bark and very large cushions of spines; cushion a round, hard mass of short, woolly hair, from which the spines—about fifty in each cushion—radiate in all directions; longest spines 2 in. or more in length; one or two new ones are developed annually, and these are bright red when young, almost black when ripe; young branches < in. to = in. in diameter. Leaves = in. apart, 3 in. to 6 in. long by 1 in. to 2 in. wide, oblong, pointed, with short petioles, and a small tuft of short, brown hair, with three or more reddish spines, in the axil of each. Flowers on the ends of the young, ripened branches, clustered in the upper leaf-axils, each flower 2 in. across, and composed of a regular circle of rosy-red petals, with a cluster of whitish stamens in the centre. They remain on the plant several weeks. Native of New Grenada. Probably *P. grandiflora* is the same as this, or a slightly different form of it. A large specimen may be obtained in a year or two by planting it in a well-drained bed of loam, in a warm, sunny house. It blossoms almost all summer if allowed to make strong growth. Pretty little flowering plants may be had by taking ripened growths from an old plant, and treating them as cuttings till rooted. In the following spring they are almost certain to produce flowers. Plants 1 ft. high, bearing a cluster of flowers, are thus annually obtained at Kew. Fig. 87 represents a short, stunted branch, probably from a specimen grown in a pot. When planted out, the leaves and spine-cushions are farther apart.

FIG. 87.—PERESKIA BLEO

P. zinniaeflora (Zinnia-flowered); Fig. 88.—Stem erect, woody, branching freely, the branches bearing oval, acuminate, fleshy, wavy-edged, green leaves, with short petioles, and a pair of spines in the axil of each. Spine-cushions on old stems crowded with stout, brown spines. Flowers rosy-red, terminal on the ripened young shoots, and composed of a whorl of broad, overlapping petals, with a cluster of stamens in the centre, the whole measuring nearly 2 in. across. This species is a native of Mexico; it grows and flowers freely if kept in a warm house.

FIG. 88. — PERESKIA ZINNIAEFLORA

CHAPTER XVII.

THE GENUS RHIPSALIS

(From *rhips*, a willow-branch; referring to the flexible, wand-like branches of some of the kinds.)

BOUT thirty species of Rhipsalis are known, most of them more peculiar than ornamental, although everyone is in some way interesting. They are remarkable for the great variety in form and habit presented by the different kinds, some of them much less resembling Cactuses than other plants. Thus, in *R. Cassytha*, the long, fleshy, whip-like branches and white berries are very similar to Mistletoe; *R. salicornoides*, with its leafless, knotty branches, resembles a Salicornia, or Marsh Samphire; another is like a Mesembryanthemum; and so on. The flowers are usually small, and composed of numerous linear sepals and petals, arranged more or less like a star, with a cluster of thin stamens in the centre, and an erect, rayed stigma. In the flat-jointed kinds, the flowers are developed singly, in notches along the margins of the young, ripened joints; in the knotted, Samphire-like kinds, they are borne on the ends of the branches; and in those with short, fleshy, leaf-like joints, they are usually placed on what appear to be flower-joints. Although the branches of these plants are usually altogether unlike the rest of the Order, yet occasionally they develop joints which are furrowed, and bear clusters of spines exactly as in the commoner forms of Cactuses.

The geographical distribution of Rhipsalis is exceptional. It is the only genus of Cactuses that has representatives in the Old World, excluding, of course, those which have been introduced by man. The bulk of the kinds of Rhipsalis occur in Central and South America, and the West Indies; but one—viz., *R. Cassytha*—is also found in Africa, Mauritius, Madagascar, and Ceylon, as well as in tropical America. Several other species are found in Madagascar, some of them only recent discoveries. The occurrence of similar or even identical plants in tropical America and Madagascar has its analogy in the Animal Kingdom as represented in the two countries.

Cultivation.—All the species appear to grow well and flower freely under cultivation, the slowest grower being, perhaps, *R. sarmentacea*. In their natural homes they are invariably found either on trees or rocks, seldom or never on the ground; but in greenhouses they may be grown in pots, a few being happiest when suspended near the glass. They do not like bright sunshine, nor should they be kept in a very shaded, moist position. There is a good collection of kinds in the Succulent-house at Kew.

Propagation.—Seeds of Rhipsalis ripen freely, and these, if sown on sandy soil, and placed on a shelf in a warm house, germinate in a few days. The development of the seedlings is exceptionally interesting, as the vegetative organs of all the kinds are very similar, and Cactus-like; the gradual transition from this character to the diverse forms which many of the species assume when mature is quite phenomenal. Cuttings will strike at almost any time, if planted in sandy soil and kept in a close, warm house till rooted. Some of the kinds thrive best when grafted on to a thin-stemmed Cereus. Treated in this way, *R. sarmentacea* makes 6 in. of growth in a season; whereas, on its own roots it would take about five years to grow as much.

The following is a selection of the species cultivated in gardens. The genus *Lepismium* is now included in Rhipsalis.

SPECIES.

R. Cassytha (derivation not known).—A pendent shrub, 4 ft. or more high, growing on rocks and the mossy trunks of trees. Branches numerous, flexuous, with small branchlets or joints springing from the ends in clusters, smooth, round, the thickness of whipcord, leafless, with numerous brown, dot-like marks scattered over the surface; under a lens these dots are seen to be tufts of very fine hairs. Flowers on the sides of the young branches, small, greenish-white, short-lived; they are developed in September, and are succeeded by white berries, exactly like those of the Mistletoe, whence the name Mistletoe Cactus, by which this species is known. An interesting and easily-grown warm greenhouse plant, native of tropical America, Africa, &c. It was introduced in 1758.

R. commune (common); Bot. Mag. 3763.—Stem straggling, branching freely, growing to a length of several feet. Branches jointed; joints varying in length, triangular, the angles compressed, and notched along the margins; notches regular, and bearing tufts of whitish hair. Strong plants produce joints over 1 in. in width. Flowers white, tinged with purple, springing singly from the notches, and composed of eight to twelve sepals and petals. Stamens and stigma erect, white, the latter four-rayed. This species is a native of Brazil, and was introduced in 1830; Flowering-season, October to December. It may be grown in a warm greenhouse, and treated as a basket-plant or as a small pot-shrub. Syn. *Lepismium commune*.

R. crispata (curled).—Stem branching freely. Branches jointed and flat, like Epiphyllum. Margins of joints notched, and slightly curled. Flowers small, white, produced singly, in November and December, in the notches on the younger joints. Fruits white, pea-like, rather rarely ripened. A free-growing, compact stove shrub, with a bright green, healthy appearance. The similarity of its branches to Epiphyllum led to its being included in that genus by Haworth.

R. c. purpurea (purple).—This variety has larger, broader joints, which are bronzy-purple in colour.

R. fasciculata (cluster-branched); Bot. Mag. 3079.—Stems terete, as thick as a goose-quill. Branches usually in clusters, and sometimes jointed, green, with small red dots and little tufts of fine, hair-like bristles. Flowers white, produced in March, springing irregularly from the older branches, small, star-like. Fruit a white berry. From its habit of growing on trees, and the character of its stems and fruit, this plant has been called parasitical. It is, however, only indebted to the tree on which is grows for moisture, for it thrives if planted in a pot or basket in ordinary soil, and kept in a stove temperature. It is a native of Brazil, and was introduced in 1831.

R. floccosa (woolly).—Stems as in *R. Cassytha*, but thicker, longer, and with the branchlets in compact clusters on the ends of the long, arching branches. The dots marking the position of the microscopic hair-tufts are in small depressions. Flowers and fruit as in *R. Cassytha*, of which this might reasonably be called a variety. This species requires warm-house temperature.

R. funalis (cord-like); Fig. 89.—Stem straggling, branched. Branches numerous, composed of long, terete joints, rather thicker than a goose-quill, glaucous-green, slightly roughened on the surface, with depressions for the dot-like cushions. Branchlets usually fascicled and spreading. Flowers white, produced in spring, on the sides of the young joints, 1 in. across, large for the genus. Introduced from Central America about 1830. An easily-grown plant, sturdy, rather straggling, but very free-flowering. In old specimens the branches become semi-pendulous. It grows best when kept in a warm house. Syn. *R. grandiflora*.

FIG. 89.—RHIPSALIS FUNALIS

R. Houlletii (Houllet's); Bot. Mag. 6089.—Stems long, graceful, branching freely, round and twig-like, or with broad wings, as in Phyllocactus. Winged or flattened portions notched, and bearing a flower in each notch. Flowers stalkless, with pointed, straw-coloured petals, forming a shallow cup about > in. across the top. Stamens and pistil white, with a tinge of red at the base. Flowering-season, November. Under cultivation, this Brazilian species forms a small, straggling shrub, about 3 ft. high, but in its native woods its stems are many feet long, and pendulous from the branches of trees. It may be grown in a warm house, in a pot, and its branches supported by a stake; or its lower stems may be fastened against a piece of soft fern-stem, into which its numerous stem-roots penetrate freely. In the winter it should be kept almost dry. The flowers remain fresh for several days, and are fragrant. A well-grown plant, when in flower, is an interesting and pretty object. It is the most ornamental kind.

R. Knightii (Knight's).—Stems and joints as in *R. commune*. Wings of joints usually broad, with red margins, and the hair in the notches in a dense tuft, nearly 1 in. long, pure white, and silk-like. Flowers small, white. This species, which thrives best under warm-house treatment, is a native of Brazil, and is usually grown only for its curious, Cereus-like stems. It forms a straggling plant about 1 ft. high. Syn. *Lepismium Knightii, Cereus Knightii*.

R. mesembryanthemoides (Mesembryanthemum-like); Bot. Mag. 3078.—A small, compact plant, with woody stems, densely covered with little fleshy, conical joints, resembling very closely the leaves of some of the Mesembryanthemums. They are green, with a few red dots, each bearing a very small tuft of the finest hair-like spines. The flowers are developed in March, from the sides of the small joints; they are = in. across, and yellowish-white. Fruit a small, white, round berry. Native of South America, whence it was introduced in 1831. When grown in a warm house, in a small, round, wire basket, filled with peat and sphagnum, this little Cactus forms a pretty tuft, which in the spring produces large numbers of white, star-like flowers.

R. myosurus (mouse-tailed); Bot. Mag. 3755.—Stems dependent, several feet long, branching freely, jointed, with three or four angles or wings; the angles flattened, reddish, notched in the margin, and bearing a tuft of white, silky hairs in each notch. Flowers small, yellow, tinged with red, springing from the notches; produced in July. Fruit not seen. A native of Brazil; introduced in 1839. This species resembles some of the angular-stemmed kinds of Cereus. It grows freely and flowers annually, if planted in a basket of fibrous soil, and suspended near the glass in a warm greenhouse or stove. It is attractive even when not in flower, owing to the form of its stems and the tufts of long, silky, white hair which spring from the notches. Syn. *Lepismium myosurus*.

R. pachyptera (thick-winged); Bot. Mag. 2820.—Stem woody; branches jointed, flattened as in Phyllocactus, with deep notches; width of joints, 2 in. or more. Flowers small, yellowish-white, borne singly in the notches in November. Fruit a small, white berry, rarely ripened. A sturdy, comparatively uninteresting stove plant, introduced from Brazil in 1830. Syn. *Cactus alatus*.

R. paradoxa (paradoxical).—Stems trailing, with numerous long branches of most extraordinary form. Imagine a three-angled, fleshy branch, often several feet in length, the angles winged, about = in. deep, green, with smooth, reddish margins. At intervals of about 2 in. the branch has the appearance of having been twisted half round. There is no other plant with branches anything like these. Flowers produced in November, in the apex of the interrupted angles, small, white. Fruit seldom ripened. A native of Brazil, whence it was introduced in 1837. There is a fine example of this trained along a rafter in the Succulent-house at Kew. The numerous branches hang down several feet from the rafter, and have a most extraordinary appearance. This species requires stove treatment.

R. penduliflora (pendulous-flowered).—A small, thin-stemmed plant, with smooth, green branches, no thicker than whipcord, and numerous fascicled or clustered, small joints, = in. long, green, with red dots, angular when young. Flowers on the tips of the terminal joints, pale yellow, = in. across, developing in August. Fruit white, Mistletoe-like. This species was introduced from tropical America in 1877, and requires stove treatment.

R. p. laxa (loose).—This variety has the branches curving, and more pendulous; in other respects it resembles the type, and requires the same treatment.

R. pentaptera (five-winged).—Stems erect; branches stiff, long-jointed, with five wing-like angles, slightly spiral, the angles notched at intervals of 1 in. Flowers in the notches, = in. across, white, produced in August. Fruit a white, Mistletoe-like berry. A curious plant from Brazil, and introduced in 1836. In stove temperature it forms a compact pot-shrub, 2 ft. high, and is worth growing on account of its singular stems.

R. rhombea (diamond-branched).—Stems and branches as in *R. crispata*, but without the wavy margins, and with more elongated joints. Flowers small, white, produced in the notches of the joints in November. Fruit a shining, milk-white berry. A compact plant from Brazil, worth growing for its bright green, leaf-like stems. It should be grown in pots, in stove temperature, and encouraged to form a globose bush.

R. Saglionis (Saglio's); Bot. Mag. 4039.—A tiny plant, similar in habit to *R. penduliflora*, but with brown branches, the small joints angled, and bearing silky hairs. The branches and joints are set at zigzag angles. Flowers pale yellow, produced in autumn on the younger joints. Fruits white, Mistletoe-like. A small, delicate plant from Buenos Ayres, not more than 6 in. high. This species requires stove treatment.

R. salicornoides (Glasswort-like); Bot. Mag. 2461.—Stem woody when old, brown, jointed like hens' toes, not quite as thick as a goose-quill. Branches in clusters; joints = in. to 1 in. long, the lower half much thinner than the upper, so that the joints look like a number of superposed, miniature clubs. Flowers pretty, on the ends of the terminal joints, yellow, becoming red with age. An erect plant, 3 ft. or more high, introduced from Brazil in 1830. The joints are clustered on the upper part of the stem. When in flower in spring this is an attractive and very remarkable-looking plant. It thrives best in stove temperature.

R. s. stricta (straight).—This variety has the joints all pointing upwards, and is much more compact than the type.

R. sarmentacea (runner-stemmed); Fig. 90.—A creeping, prostrate plant, with round stems as thick as a goose-quill, and attaching themselves to tree-trunks or other bodies by means of numerous adventitious roots, which spring from the under side of the stems. Surface of stem furrowed, and covered with numerous small clusters of short, hair-like, whitish spines. Flowers 1 in. across, springing from the sides of the stems, with pointed, creamy-white petals; stamens spreading; stigma erect, four-lobed. Fruit small, currant-like. This is a pretty little species, introduced from Brazil in 1858; it is, however, a very slow grower, plants ten years old being only a few inches in diameter. It should be grown in stove temperature, in a basket of peat fibre, or, better still, on a piece of soft fern-stem. It is always found on the branches or trunks of trees when growing wild.

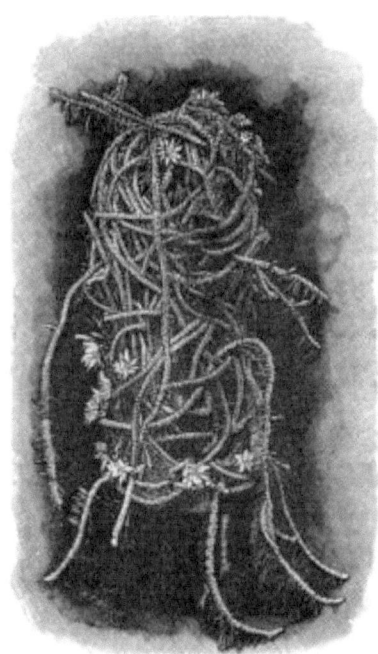

FIG. 90.—RHIPSALIS SARMENTACEA

R. Swartziana (Swartz's).—Older stems three-angled, young ones flattened, jointed; joints 2 in. broad, stiff with deep notches. Flowers in the notches, small, white, produced in June. This species is a native of Jamaica, and was introduced in 1810. A stiff, ungraceful plant, about 2 ft. high, very similar in its branches to a Phyllocactus. This species requires the temperature of a stove.

R. trigona (triangular).—Habit straggling; branches usually in forks, < in. in diameter, three-angled; angles wavy or slightly notched, grey-green. Flowers small, produced in spring in the notches of the angles, white. Fruit a white berry. A thin, Brazilian plant, not unlike a Lepismium, but without the silky hairs in the notches of the angles. This species also requires to be grown in stove temperature.

CHAPTER XVIII.

TEMPERATURES.

O enable growers to make a selection of species according to the accommodation that can be afforded for Cactuses, all that are described in this book are here classified in three groups: (1) Species which thrive in a cool-house or frame; (2) Species which can only be successfully grown in a warm house or stove; and (3) Species which are hardy in the more favoured portions of the United Kingdom.

COOL-HOUSE OR FRAME.

Temperature: Summer, that of the open air.
Temperature: Winter — day, 50 deg. to 60 deg.; night, 40 deg. to 45 deg.

Cereus Berlandieri.
Cereus Blankii.
Cereus caespitosus.
Cereus cirrhiferus.

Cereus speciosissimus.
Echinocactus brevihamatus.
Echinocactus centeterius.
Echinocactus cinnabarinus.

Cereus ctenoides.
Cereus enneacanthus.
Cereus flagelliformis.
Cereus Leeanus.
Cereus leptacanthus.
Cereus multiplex.
Cereus paucispinus.
Cereus pentalophus.
Cereus polyacanthus.
Cereus procumbens.
Cereus reductus.

Echinocactus Cummingii.
Echinocactus echidne.
Echinocactus Emoryi.
Echinocactus gibbosus.
Echinocactus hexaedrophorus.
Echinocactus Leeanus.
Echinocactus Mackieanus.
Echinocactus mamillarioides.
Echinocactus rhodophthalmus.
Echinocactus texensis.
Echinocactus uncinatus.

Echinocactus viridescens.
Echinopsis, all the kinds.
Mamillaria atrata.
Mamillaria chlorantha.
Mamillaria dasyacantha.
Mamillaria elegans.
Mamillaria elephantidens.
Mamillaria elongata.
Mamillaria fissurata.
Mamillaria floribunda.
Mamillaria gracilis.
Mamillaria longimamma.
Mamillaria macromeris.
Mamillaria macrothele.
Mamillaria multiceps.
Mamillaria phellosperma.
Mamillaria Schelhasii.
Mamillaria Schiedeana.

Mamillaria semperviva.
Mamillaria stella-aurata.
Mamillaria tuberculosa.
Mamillaria uncinata.
Mamillaria viridis.
Mamillaria Wildiana.
Mamillaria Zuccchariniana.
Opuntia arborescens.
Opuntia aurantiaca.
Opuntia corrugata.
Opuntia cylindrica.
Opuntia Davisii.
Opuntia Engelmanni.
Opuntia Ficus-indica.
Opuntia hystricina.
Opuntia macrorhiza.
Opuntia occidentalis.

WARM-HOUSE OR STOVE.

It is possible that some of those included in this division will eventually prove hardier than is at present supposed. Many of the species now classed as cool-house plants, and even some of those which are hardy, were formerly grown in the stove.

Temperature: Summer-day, 70 deg. to 90 deg.; night, 65 deg. to 75 deg.
Temperature: Winter-day, 60 deg. to 70 deg.; night, 55 deg. to 60 deg.

Cereus caerulescens.
Cereus extensus.
Cereus fulgidus.
Cereus giganteus.
Cereus grandiflorus.
Cereus Lemairii.
Cereus Macdonaldiae.
Cereus Mallisoni.
Cereus Napoleonis.
Cereus nycticalus.
Cereus peruvianus.
Cereus pleiogonus.
Cereus repandus.
Cereus serpentinus.
Cereus Royeni.
Cereus triangularis.
Cereus variabilis.
Echinocactus concinnus.
Echinocactus coptonogonus.
Echinocactus cornigerus.
Echinocactus corynodes.
Echinocactus cylindraceus.
Echinocactus Haynii.
Echinocactus horizonthalonis.
Echinocactus Le Contei.
Echinocactus longihamatus.
Echinocactus mamillosus.
Echinocactus multiflorus.
Echinocactus myriostigma.
Echinocactus obvallatus.
Echinocactus Ottonis.
Echinocactus pectiniferus.
Echinocactus polycephalus.
Echinocactus Pottsii.
Echinocactus scopa.
Echinocactus sinuatus.
Echinocactus tenuispinus.
Echinocactus turbiniformis.

Echinocactus Visnaga.

Echinocactus Williamsii.

Echinocactus Wislizeni.

Epiphyllum, all the kinds.

Leuchtenbergia.

Mamillaria, all not mentioned under " Cool-house Or Frame."

Melocactus, all the kinds.

Opuntia, all not included under "Cool-house or Frame."

Pelecyphora.

Pereskia, all the kinds.

Phyllocactus, all the kinds.

Pilocereus, all the kinds.

Rhipsalis, all the kinds.

OUTDOOR.

The following may be grown out of doors in the more favoured parts of England. For directions as to culture, *see* the chapter on Cultivation:—

Cereus Fendleri.	Opuntia missouriensis.
Echinocactus Simpsoni.	Opuntia Rafinesquii.
Mamillaria vivipara.	Opuntia R. arkansana.
Opuntia brachyarthra.	Opuntia vulgaris.

CHAPTER XIX.

DEALERS IN CACTUSES.

HE difficulty experienced by amateurs in procuring plants of many kinds of Cactus has suggested to us that a list of some of the principal dealers in these plants would prove of service to English growers. So far as we know, there is no nurseryman in England who makes a specialty of Cactuses. Plants of such well-known genera as *Epiphyllum, Phyllocactus,* and *Cereus* in part, may be obtained in England, but for a collection of representative kinds we must perforce apply to Continental nurserymen. The most reliable of these for Cactuses are:

GERMANY.

Messrs. HAAGE & SCHMIDT, Erfurt.
Herr F. A. HAAGE, Junior, Erfurt.
Messrs. M\LLER & SAUBER, Kassel, Hanover.
Herr H. HILDMANN, Oranienburg, Brandenburg.
Herr ERNST BERGE, Leipsic.

FRANCE.

M. EBERLE, Avenue de St. Ouen, 146, Paris.

M. JAMIN, Rue Lafontaine, 42, ` St. Ouen, Paris.

BELGIUM.

M. LOUIS DE SMET, Ledeberg, Ghent.

M. BETTES, Borgerhont, Antwerp.

M. F. VERMUELEN, Rue Van Peet, Antwerp.

AMERICA.

Mr. C. RUNGE, San Antonio, Texas.
Messrs. REASONER BROTHERS, Florida.

SPECIES INDEX

Species are listed alphabetically according to Watson's nomenclature. The name(s) that is more likely to be recognised by modern readers is listed in brackets. I have used Anderson's book—*The Cactus Family* (Timber Press, 2001)—as my main guide. Monographs by Craig and by Pilbeam were invaluable in identifying 'Mamillarias'.

* Plants illustrated in the text.

- Cereus Berlandieri (*Echinocereus berlandieri*) *
- Cereus Blankii (*Echinocereus berlandieri*) *
- Cereus caerulescens (*Cereus aethiops*)
- Cereus caespitosus (*Echinocereus reichenbachii ssp. caespitosus*) *
- Cereus cirrhiferus (*Echinocereus cinerascens*)
- Cereus ctenoides (*Echinocereus dasyacanthus*) *
- Cereus enneacanthus (*Echinocereus enneacanthus*)*
- Cereus extensus (*Selenicereus sp.?*)
- Cereus Fendleri (*Echinocereus fendleri*)
- Cereus flagelliformis (*Aporocactus (Disocactus) flagelliformis*)
- Cereus fulgidus (*Gymnocalycium gibbosum*)
- Cereus giganteus (*Carnegiea gigantea*) *
- Cereus grandiflorus (*Selenicereus grandiflorus*)
- Cereus Leeanus (*Echinocereus polyacanthus*)
- Cereus Lemairii (*Hylocereus lemairei*)
- Cereus leptacanthus (*Echinocereus pentalophus*)*
- Cereus Macdonaldiae (*Selenicereus macdonaldiae*)
- Cereus Mallisoni (X *Helioporus smithii*)
- Cereus multiplex (*Echinopsis oxygona*) *
- Cereus multiplex cristatus (*Echinopsis oxygona fa. cristata*) *
- Cereus Napoleonis (*Hylocereus trigonus*)
- Cereus nycticalus (*Selenicereus pteranthus*) *
- Cereus paucispinus (*Echinocereus coccineus ssp. paucispinus*)
- Cereus pentalophus (*Echinocereus pentalophus*)
- Cereus peruvianus (*Cereus repandus*)

- Cereus pleiogonus (*Echinocereus sp.* — no longer identifiable)*
- Cereus polyacanthus (*Echinocereus polyacanthus*)
- Cereus procumbens (*Echinocereus pentalophus ssp. procumbens*) *
- Cereus reductus (Hybrid with *Selenicereus sp.* as one probable parent)
- Cereus repandus (*Cereus repandus*) *
- Cereus Royeni (*Pilosocereus royenii*)
- Cereus serpentinus (*Peniocereus (Nyctocereus) serpentinus*) *
- Cereus speciosissimus (*Disocactus speciosus*)
- Cereus triangularis (*Hylocereus triangularis*)
- Cereus variabilis (*Acanthocereus tetragonus*)

- Echinocactus brevihamatus (*Parodia (Notocactus) alacriportana ssp. brevihamata*)
- Echinocactus centeterius (*Eriosyce (Neoporteria) curvispina* — possibly?)
- Echinocactus cinnabarinus (*Echinopsis (Lobivia) cinnabarina*)
- Echinocactus concinnus (*Parodia (Notocactus) concinna*) *
- Echinocactus coptonogonus (*Stenocactus (Echinofossulocactus) coptonogonus*) *
- Echinocactus cornigerus (*Ferocactus latispinus*) *
- Echinocactus corynodes (*Parodia (Notocactus) sellowii*) *
- Echinocactus crispatus (Stenocactus (Echinofossulocactus) crispatus) *
- Echinocactus Cummingii (*Rebutia (Weingartia) neocummingii*)
- Echinocactus cylindraceus (*Ferocactus cyclindraceus (acanthodes)*)
- Echinocactus echidne (*Ferocactus echidne*)
- Echinocactus Emoryi (*Ferocactus emoryi*) *
- Echinocactus gibbosus (*Gymnocalycium gibbosus*)
- Echinocactus Haynii (*Matucana haynei*) *
- Echinocactus hexaedrophorus (*Thelocactus hexaedrophorus*) *

- Echinocactus horizonthalonis (*Echinocactus horizonthalonius*) *
- Echinocactus Le Contei (*Ferocactus cylindraceus (acanthodes) ssp. lecontei*) *
- Echinocactus Leeanus (*Gymnocalycium leeanum*)
- Echinocactus longihamatus (*Ferocactus hamatacanthus*) *
- Echinocactus Mackieanus (*Gymnocalycium mackieanum*)
- Echinocactus mamillarioides (*Eriosyce (Neoporteria) curvispina* – possibly?)
- Echinocactus mamillosus (*Echinopsis mamillosa*)
- Echinocactus multiflorus (*Gymnocalycium monvillei*)
- Echinocactus myriostigma (*Astrophytum myriostigma*) *
- Echinocactus obvallatus (*Stenocactus (Echinofossulocactus) obvallatus*) *
- Echinocactus Ottonis (*Parodia (Notocactus) ottonis*)
- Echinocactus pectiniferus (probably *Echinocereus pectinatus*)
- Echinocactus polycephalus (*Echinocactus polycephalus*)*
- Echinocactus Pottsii (*Ferocactus pottsii*)
- Echinocactus rhodophthalmus (*Thelocactus bicolor*)
- Echinocactus scopa (*Parodia (Notocactus) scopa*) *
- Echinocactus scopa cristata (*Parodia (Notocactus) scopa fa. cristata*) *
- Echinocactus Simpsonii (*Pediocactus simpsonii*)
- Echinocactus sinuatus (*Ferocactus hamatacanthus ssp. sinuatus*)
- Echinocactus tenuispinus (*Parodia (Notocactus) ottonis*)
- Echinocactus texensis (*Echinocactus texensis*) *
- Echinocactus turbiniformis (*Strombocactus disciformis*)
- Echinocactus uncinatus (*Sclerocactus uncinatus*) *
- Echinocactus viridescens (*Ferocactus viridescens*)
- Echinocactus visnaga (*Echinocactus platyacanthus*) *
- Echinocactus Williamsii (*Lophophora williamsii*)
- Echinocactus Wislizenii (*Ferocactus wislizenii*) *

- Echinopsis campylacantha (*Echinopsis leucantha*)
- Echinopsis cristata (*Echinopsis obrepanda*)

- Echinopsis cristata purpurea (*Echinopsis obrepanda v. purprea*)
- Echinopsis Decaisneanus (identification now uncertain) *
- Echinopsis Eyriesii (*Echinopsis eyriesii*)
- Echinopsis Eyriesii flore-pleno (*Echinopsis eyriesii*) *
- Echinopsis Eyriesii glauca (*Echinopsis eyriesii*)
- Echinopsis oxygonus (*Echinopsis oxygona*)
- Echinopsis Pentlandi (*Echinopsis (Lobivia) pentlandii*) *
- Echinopsis Pentlandi longispinus (*Echinopsis (Lobivia) pentlandii*) *
- Echinopsis tubiflorus (*Echinopsis tubiflora*)

- Epiphyllum Russellianum (*Schlumbergera russelliana*) *
- Epiphyllum truncatum (*Schlumbergera truncata*)

- Leuchtenbergia principis (*Leuchtenbergia principis*) *

- Mamillaria angularis (*Mammillaria compressa*)
- Mamillaria applanata (*Mammillaria heyderi ssp. hemisphaerica*)
- Mamillaria atrata (*Eriosyce (Neoporteria) subgibbosa*)
- Mamillaria bicolor (*Mammillaria geminispina*)
- Mamillaria chlorantha (*Escobaria deserti*)
- Mamillaria cirrhifera (*Mammillaria compressa*)
- Mamillaria clava (*Coryphantha clava*)
- Mamillaria dasyacantha (*Escobaria dasyacantha*)
- Mamillaria discolor (*Mammillaria discolor*)
- Mamillaria dolichocentra (*Mammillaria polythele*) *
- Mamillaria echinata (*Mammillaria elongata ssp. echinaria*)
- Mamillaria echinus (*Coryphantha echinus*) *
- Mamillaria elegans (*Mammillaria haageana (elegans)*)
- Mamillaria elephantidens (*Coryphantha elephantidens*) *
- Mamillaria elongata (*Mammillaria elongata*)
- Mamillaria fissurata (*Ariocarpus fissuratus*) *
- Mamillaria floribunda (*Eriosyce (Neoporteria) subgibbosa*)
- Mamillaria gracilis (*Mammillaria vetula ssp. gracilis*)

- Mamillaria Grahami (*Mammillaria grahamii*)
- Mamillaria Haageana (*Mammillaria haageana*) *
- Mamillaria longimamma (*Mammillaria longimamma*) *
- Mamillaria macromeris (*Coryphantha macromeris*) *
- Mamillaria macrothele (*Coryphantha octacantha*)
- Mamillaria micromeris (*Epithelantha micromeris*) *
- Mamillaria multiceps (*Mammillaria prolifera ssp. texana*)
- Mamillaria Neumanniana (*Mammillaria magnimamma*)
- Mamillaria Ottonis (*Coryphantha ottonis*)
- Mamillaria pectinata (*Mammillaria pectinifera*) *
- Mamillaria phellosperma (*Mammillaria tetrancistra*)
- Mamillaria pulchra (*Mammillaria rhodantha*)
- Mamillaria pusilla (*Mammillaria prolifera*)
- Mamillaria pycnacantha (*Coryphantha pycnacantha*)
- Mamillaria sanguinea (*Mammillaria spinosissima*) *
- Mamillaria Scheerii (*Coryphantha poselgeriana*)
- Mamillaria Schelhasii (*Mammillaria crinita*) *
- Mamillaria Schiedeana (*Mammillaria schiedeana*)
- Mamillaria semperviva (*Mammillaria sempervivi*) *
- Mamillaria senilis (*Mammillaria senilis*)
- Mamillaria stella-aurata (*Mammillaria elongata*)
- Mamillaria sub-polyhedra (*Mammillaria polyedra*) *
- Mamillaria sulcolanata (*Coryphantha sulcolanata*)
- Mamillaria tetracantha (*Mammillaria polythele*)
- Mamillaria tuberculosa (*Escobaria tuberculosa*)
- Mamillaria turbinata (*Strombocactus disciformis* ?)
- Mamillaria uncinata (*Mammillaria uncinata*)
- Mamillaria vetula (*Mammillaria vetula*)
- Mamillaria villifera (*Mammillaria polyedra*)
- Mamillaria viridis (*Mammillaria karwinskiana*)
- Mamillaria vivipara (*Escobaria vivipara*)
- Mamillaria vivipara v. radiosa (*Escobaria vivipara*) *
- Mamillaria Wildiana (*Mammillaria crinita ssp. wildii*)
- Mamillaria Wrightii (*Mammillaria wrightii*)
- Mamillaria Zucchariniana (*Mammillaria magnimamma*)
- Melocactus communis (*Melocactus intortus*) *

- Melocactus depressus (*Melocactus violaceus*)
- Melocactus Miquelii (*Melocactus intortus*) *

- Opuntia arborescens (*Cylindropuntia imbricata*)
- Opuntia arbuscula (*Cylindropuntia arbuscula*)
- Opuntia arenaria (*Opuntia polyacantha v. arenaria*)
- Opuntia Auberi (*Opuntia auberi*)
- Opuntia aurantiaca (*Opuntia aurantiaca*)
- Opuntia basilaris (*Opuntia basilaris*) *
- Opuntia Bigelovii (*Cylindropuntia bigelovii*)
- Opuntia boliviana (*Cumulopuntia boliviana*) *
- Opuntia brachyarthra (*Opuntia fragilis*) *
- Opuntia braziliensis (*Brasiliopuntia brasiliensis*)
- Opuntia candelabriformis (*Opuntia spinulifera*)
- Opuntia clavata (*Grusonia clavata*)
- Opuntia cochinellifera (*Opuntia cochenillifera*)
- Opuntia corrugata (*Tunilla corrugata*)
- Opuntia curassavica (*Opuntia curassavica*)
- Opuntia cylindrica (*Austrocylindropuntia cylindrica*)
- Opuntia cylindrica cristata (*Austrocylindropuntia cylindrica fa. cristata*)
- Opuntia Davisii (*Cylindropuntia davisii*)
- Opuntia decumana (*Opuntia ficus-indica*)
- Opuntia diademata (*Tephrocactus articulatus*)
- Opuntia Dillenii (*Opuntia dillenii*) *
- Opuntia echinocarpa (*Cylindropuntia echinocarpa*)
- Opuntia Emoryi (*Grusonia emoryi*)
- Opuntia Engelmanni (*Opuntia engelmannii*)
- Opuntia Ficus-indica (*Opuntia ficus-indica*) *
- Opuntia filipendula (*Opuntia macrorhiza v. pottsii*) *
- Opuntia frutescens (*Cylindropuntia leptocaulis*)
- Opuntia Grahami (*Grusonia grahamii*)
- Opuntia horrida (*Opuntia tuna*)
- Opuntia hystricina (*Opuntia polyacantha v. hystricina*)
- Opuntia leptocaulis (*Cylindropuntia leptocaulis*)
- Opuntia leucotricha (*Opuntia leucotricha*)

- Opuntia macrocentra (*Opuntia macrocentra*)
- Opuntia macrorhiza (*Opuntia macrorhiza*) *
- Opuntia microdasys (*Opuntia microdasys*)
- Opuntia missouriensis (*Opuntia polyacantha*)
- Opuntia monacantha (*Opuntia monacantha*)
- Opuntia nigricans (*Opuntia elatior*)
- Opuntia occidentalis (*Opuntia X occidentalis*)
- Opuntia Parmentieri (*Opuntia longispina var. brevispina*?)
- Opuntia Parryi (*Cylindropuntia californica*)
- Opuntia Rafinesquii (*Opuntia humifusa*) *
- Opuntia rosea (*Cylindropuntia rosea*) *
- Opuntia Salmiana (*Opuntia salmiana*)
- Opuntia spinosissima (*Consolea spinosissima*)
- Opuntia subulata (*Austrocylindropuntia subulata*)
- Opuntia Tuna (*Opuntia tuna*) *
- Opuntia tunicata (*Cylindropuntia tunicata*)
- Opuntia vulgaris (*Opuntia ficus-indica*)
- Opuntia Whipplei (*Cylindropuntia whipplei*)

- Pelecyphora aselliformis (*Pelecyphora aselliformis*) *
- Pereskia aculeata (*Pereskia aculeata*)
- Pereskia aculeata rubescens (*Pereskia aculeata*)
- Pereskia Bleo (*Pereskia bleo*) *
- Pereskia zinniaeflora (*Pereskia zinniiflora*) *

- Phyllocactus Ackermannii (*Disocactus ackermannii* or hybrid) *
- Phyllocactus anguliger (*Epiphyllum anguliger*) *
- Phyllocactus biformis (*Disocactus biformis*) *
- Phyllocactus crenatus (*Epiphyllum crenatum*)
- Phyllocactus grandis (*Epiphyllum oxypetalum*)
- Phyllocactus Hookeri (*Epiphyllum hookeri*)
- Phyllocactus latifrons (*Epiphyllum oxypetalum*)
- Phyllocactus phyllanthus (*Epiphyllum phyllanthus*)
- Phyllocactus phyllanthoides (*Disocactus phyllanthoides*)

- Pilocereus Br|nnonii (*Oreocereus celsianus*) *
- Pilocereus Houlletianus (*Pilosocereus leucocephalus*) *
- Pilocereus senilis (*Cephalocereus senilis*)

- Rhipsalis Cassytha (*Rhipsalis baccifera*)
- Rhipsalis commune (*Lepismium cruciforme*)
- Rhipsalis crispata (*Rhipsalis crispata*)
- Rhipsalis crispata purpurea (*Rhipsalis crispata*)
- Rhipsalis fasciculata (*Rhipsalis baccifera*)
- Rhipsalis floccosa (*Rhipsalis floccosa*)
- Rhipsalis funalis (*Rhipsalis grandiflora*) *
- Rhipsalis Houlletii (*Lepismium houlletianum*)
- Rhipsalis Knightii (*Lepismium cruciforme*)
- Rhipsalis mesembryanthemoides (*Rhipsalis mesembryanthemoides*)
- Rhipsalis myosurus (*Lepismium cruciforme*)
- Rhipsalis pachyptera (*Rhipsalis pachyptera*)
- Rhipsalis paradoxa (*Rhipsalis paradoxa*)
- Rhipsalis penduliflora (*Rhipsalis cereuscula*)
- Rhipsalis penduliflora laxa (*Rhipsalis cereuscula*)
- Rhipsalis pentaptera (*Rhipsalis pentaptera*)
- Rhipsalis rhombea (identification now uncertain)
- Rhipsalis Saglionis (*Rhipsalis cereuscula*)
- Rhipsalis salicornoides (*Hatiora salicornioides*)
- Rhipsalis salicornoides stricta (*Hatiora salicornioides*)
- Rhipsalis sarmentacea (*Lepismium lumbricoides*) *
- Rhipsalis Swartziana (*Pseudorhipsalis alata*)
- Rhipsalis trigona (*Rhipsalis trigona*)

Back to Contents list